PETRIFIED LIGHTNING

PETRIFIED LIGHTNING

And More Amazing Stories from "Our Fascinating Earth"

PHILIP SEFF, PH.D., AND NANCY R. SEFF, M.ED.

CONTEMPORARY BOOKS
A TRIBUNE COMPANY

Library of Congress Cataloging-in-Publication Data

Seff, Philip.
    Petrified lightning : and more amazing stories from "Our fascinating earth" / Philip Seff and Nancy R. Seff.
       p.   cm.
    ISBN 0-8092-3250-2
    1. Science—Miscellanea.   2. Natural history—Miscellanea.   3. Curiosities and wonders—Miscellanea.   I. Seff, Philip. Our fascinating earth.   II. Seff, Nancy R.   III. Title.
Q173.S444   1996
500—dc20   96-11465
                                                                     CIP

Interior illustrations by Mel Chadwick and Doug Wellons
Front cover illustration by Mark Anderson

Copyright © 1996 by Philip Seff and Nancy R. Seff
All rights reserved
Published by Contemporary Books
An imprint of NTC/Contemporary Publishing Company
Two Prudential Plaza, Chicago, Illinois 60601-6790
Manufactured in the United States of America
International Standard Book Number: 0-8092-3250-2
10 9 8 7 6 5 4 3 2 1

Dedicated to
the *Redlands Daily Facts*,
where it all began

CONTENTS

Acknowledgments   XVII
Introduction   XIX

Chapter One

A Dinosaur Classic   1
(PALEONTOLOGY)

Erotic Buzzards   6
(ORNITHOLOGY)

A Medal of Honor   8
(ARCHAEOLOGY/ZOOLOGY)

How Not to Find a Volocano   10
(GEOLOGY)

Fish Out of Water   11
(ICHTHYOLOGY)

The Real Middle of Nowhere   15
(NATURAL HISTORY)

The Golden Age of Medical Quackery (MEDICINE)	15
Lost: One Ungrateful Slave (ARCHAEOLOGY)	19
The Unsuicidal Lemming (ZOOLOGY)	20
Fossil Forests of Yellowstone (PALEONTOLOGY)	23
Formula for Survival (ENTOMOLOGY)	24
The Song of Spring (ORNITHOLOGY)	27
Canine Sky Diver (ZOOLOGY)	28
Kamikaze: Divine Wind (METEOROLOGY [OCEAN])	29
Monster Born in a Dream (LEGEND)	33

Chapter Two

The Great Uranium Rush (MINERALOGY)	37
A Harness for the Wind (ARCHAEOLOGY)	40
King Cobra (HERPETOLOGY)	41
The Year of the Popes (MEDICINE)	44
Firestorms (METEOROLOGY)	45
The Ultimate Conversation Piece (PALEONTOLOGY)	48
Isn't Love Grand? (ZOOLOGY/ENTOMOLOGY)	49

Seismic Cows (GEOLOGY)	55
Land of the Roc (LEGEND/ORNITHOLOGY)	55
Don't Try This at Home (MINERALOGY)	58
Stone Age Baseball (ANTHROPOLOGY)	58
Pig Soldiers (ARCHAEOLOGY)	61
Shark Myths Dispelled (ICHTHYOLOGY)	62
Paleolithic Kitchen Rejects (ANTHROPOLOGY)	66
Today's Oldest Living Things (BACTERIOLOGY)	67

Chapter Three

Flu, 1918–1919 (MEDICINE)	69
When One Sex Prevails (HERPETOLOGY)	74
The Tainted Blue (MINERALOGY)	75
Kosher Oil (ARCHAEOLOGY)	77
It Blends with the Snow (ZOOLOGY)	78
Ancient Fossil Fuel (ANTHROPOLOGY)	82
The Sky for a Ceiling (METEOROLOGY)	83
The Hanging That Prompted a Library (LEGEND)	84

Then Comes the Undertaker Bee (Entomology)	86
Catch a Falling Star (Astronomy)	88
The Day the Mountains Walked (Geology)	91
"Who Murdered the Veterans?" (Meteorology)	92
Herding Tendencies of Dinosaurs (Paleontology)	93
The Terrible Bird (Ornithology)	93

Chapter Four

Petrified Lightning (Meteorology)	99
Shark Pen (Ichthyology)	102
Prehistoric Pit Barbecue (Anthropology)	104
The Great Mouse War (Zoology)	106
Life, Life Everywhere (Bacteriology)	108
Cloak of Darkness (Zoology)	110
The Little Ice Age (Geology)	111
Pie-Eyed Elephants and Other "Substance Abusers" (Zoology)	116
Riches from the Sea (Mineralogy [Ocean])	118
The Windiest Place in the World (Meteorology)	120

Floating Water   120
(Hydrology)

Birth of a Legend   122
(Legend)

A Nose Is a Nose   124
(Zoology)

Chapter Five

Paleopathology   127
(Paleontology)

Never Insult a Priest   131
(Psychology)

Lion Trouble   132
(Zoology)

The Naming of the Horse Latitudes   134
(Oceanography)

Oology   135
(Ornithology)

Robinson Crusoe, a Scottish Pirate   137
(Legend)

The Rounding of the Earth   137
(Geology)

So Far, yet So Near   140
(Meteorology)

Forever Beautiful   141
(Archaeology)

How England's Plague Freed America   143
(Pathology)

Laziest of All   145
(Herpetology)

The Greening of the Earth   145
(Paleobotany)

Peter's Mistake   150
(Psychology)

Venus (ASTRONOMY)	151
Stone Age Rolling Stones (ANTHROPOLOGY)	154

Chapter Six

Waiting for the Big One (GEOLOGY)	157
Phantom Treasure Ship (LEGEND)	161
An Invasion of Guam (HERPETOLOGY)	162
How to Fix a Broken Arm (BOTANY)	166
Improved Birdbrains (ORNITHOLOGY)	166
A Walk in the Sun (ANTHROPOLOGY)	168
White—Not Always Right (METEOROLOGY)	174
This Evolving Atmosphere (HISTORICAL GEOLOGY)	176
They Also Serve (ZOOLOGY)	178
Dinosaur Eater (PALEONTOLOGY)	182
Who Were the Real Barbarians? (ARCHAEOLOGY)	183
The Moth and the Candle (ENTOMOLOGY)	185
Trap-Door Spider Versus the Wasp (ENTOMOLOGY)	186
Dust to Dust (GEOLOGY)	187

Chapter Seven

Herds of Dinosaurs and Bison   191
(Paleontology)

Requiem for a Hero   193
(Zoology)

Does the Menu Bug You?   195
(Entomology)

From Out of a Cloud   197
(Astronomy)

A Dream of One World   197
(Archaeology)

Shark Prank   199
(Ichthyology)

Not So Cuckoo   200
(Ornithology)

Legendary Earthquakes   202
(Geology)

Wise and Not-So-Wise Men of the East   203
(Psychology)

A Faithless Wife   205
(Ornithology)

The Most Dangerous Animals in Africa   206
(Zoology)

Ravishing Raisin   207
(Zoology)

Symbolic Barber Pole   209
(Medicine)

Rise and Fall of a Hurricane   211
(Meteorology)

More than a Mouthful   214
(Ichthyology)

Chapter Eight

A Fraud Called *Missourium* (PALEONTOLOGY)	217
The Hoax That Was Real (ZOOLOGY)	220
Mesozoic Lifestyles (PALEONTOLOGY)	221
The World's Worst Weather (METEOROLOGY)	225
Termite Soldiers (ENTOMOLOGY)	226
The Slave and the Milkmaid (MEDICINE)	227
Creatures That Time Forgot: Living Fossils (BIOLOGY)	228
Human Counterparts (ZOOLOGY)	232
Heavy Nest (ORNITHOLOGY)	234
Planet of the Insects (ENTOMOLOGY)	235
Mars—an Unresortful Planet (ASTRONOMY)	236
"Mighty Mouse" (HERPETOLOGY)	239
The Fish That Almost Changed History (HISTORY/ICHTHYOLOGY)	240
Bear Hunt, Stone Age Style (ANTHROPOLOGY)	242

Chapter Nine

Gold Fever (MINERALOGY)	243

"How Doth the Little Crocodile" (HERPETOLOGY)	248
War Stories (HISTORY)	252
Adaptation for Survival (PHYSIOLOGY)	255
Insect Jewelry (ENTOMOLOGY)	258
Written in the Rocks of Time (PALEONTOLOGY)	258
Ostrich Massacre (ORNITHOLOGY)	261
Tropical Antarctica (GEOLOGY)	262
Hairy Apes and Humans (ZOOLOGY)	264
Conquest Through Religion (ARCHAEOLOGY)	264
Galveston's Day of Infamy (METEOROLOGY)	268
BOINGGG! (ENTOMOLOGY)	273
Neanderthal Bedouins (ANTHROPOLOGY)	274
The Time Machine (ASTRONOMY)	274

Chapter Ten

"The Tempest" (METEOROLOGY)	277
Sit-Down Strike (ARCHAEOLOGY)	279
Animal Forecasters (GEOLOGY)	282

A Most Dangerous Creature   284
    (Entomology)

Strange Matrimonial Rituals   287
    (Legends)

Weird Remedies   288
    (Legends)

A Giant of a Kangaroo   290
    (Zoology)

The World Herd   290
    (Zoology)

Hazards of Being a Bird   293
    (Ornithology)

Black as a Diamond   295
    (Mineralogy)

National Drink of Ancient Egypt   296
    (Archeology)

Women in Early Medicine   296
    (Medicine)

Green Polar Bears   297
    (Zoology)

In Unity Is Strength   298
    (Zoology)

Coffin for the Dinosaur   300
    (Paleontology)

Bibliography   309
Index   323

Acknowledgments

Adequate acknowledgment of all the assistance we have received in the writing of this book is a difficult task. Initial encouragement came from readers of our first volume who wanted more stories about this fascinating earth. We have been attentive to the opinions and suggestions of reviewers, other writers, scholars in a variety of disciplines, and especially our readers.

We gratefully acknowledge particular assistance from the following people:

Rick Right, who spent many hours reviewing the manuscript and offered numerous suggestions, many of which we accepted, and Lisa Gold, whose list of favorite essays in the manuscript was, fortunately, in direct contrast with ours.

Dr. Ronald Seff and Harriet Garfinkle, who together excised clichés, overly dramatic phrases, and other excesses, reducing reader distractions to a minimum.

Dr. Judson Sanderson, who applied his mathematical expertise to improve the clarity and readability of the text.

Dan Anderson, whose ability to reduce convoluted concepts to simple statements has made our book more accessible to all readers, and Fran Anderson, the generic avid reader who knows what readers want.

Among unnamed sources of information are people, only a phone call away, who answered our queries, sent requested material, and referred us to other wellsprings of information. Nor can we ignore the many acquaintances whose offhand remarks may have become part of the text.

Introduction

The second volume of *Our Fascinating Earth* is a continuation of the first, which was originally published in 1990. Both volumes are by-products of the illustrated panel of the same name that has appeared in newspapers nationwide for over 15 years. As did the first volume, the second contains a mosaic of stories and incidents that explore a broad range of scientific disciplines. Some articles are lengthy, others anecdotal.

Although the arrangement of articles appears to be arbitrary, the random order is intentional. As readers sample the book for something in line with their specific interests, their attention is often drawn to a new topic, and their range of interests grows wider. To simplify the search for a favorite discipline, the table of contents identifies the broad subjects of science.

Many articles are longer and more detailed than those in the earlier volume. The intent is not to cram more information into a few pages but to offer evidence and examples that support the fact or idea being developed and to add drama and substance.

We anticipate that this volume, like the first, will appeal to readers across a broad spectrum of ages and interests. Our strategy for both books has been to present science in an interesting, painless, entertaining manner while preserving its accuracy and instructional value.

Presuming that many readers will be familiar with the first volume, we avoid repeating information except as brief joggers of memory.

With the planet and the universe beyond as our source for material, we have never needed to stretch facts or to dredge beyond the wonders of the natural world to find truths stranger than fiction. The subject matter is endless and ever changing, and we need only to look more closely and wonder more intently at the mysterious world around us in order to share our growing awareness of how nature works.

Although scientific books and periodicals provide the sources for most of our information, many of the anecdotes are personal, occasionally even eyewitness accounts. Geologist Philip Seff has been on many paleontological, archaeological, and mineralogical expeditions. His firsthand experiences at exploration and discovery enable him to breathe life into many of the articles. Nancy Seff, teacher and librarian, is a researcher with unlimited resourcefulness in exhuming facts and supporting evidence.

In volume one of *Our Fascinating Earth* we sought to make people more aware and appreciative of the world around them. But the growing environmental concern of the 1990s impels us to assume a more active role. We continue to observe the many facets of the natural world and to marvel at how efficiently they harmonize into a symphony of survival. We realize humans may be the only creatures that can comprehend the total picture, not only of these relationships between plants and animals and biospheres but also of the present as a key to the past. We hope to reawaken our readers to their unity with the cosmos and to their responsibility in conserving the natural world and sharing it with the entire community of living things. If readers become more aware of their oneness with the natural universe, we will have accomplished a worthwhile goal. Together all of us may contribute to the preservation of some particle of the eternal symphony of nature for the generations that follow.

Philip Seff
Nancy Seff

PETRIFIED LIGHTNING

CHAPTER ONE

A Dinosaur Classic

Of the more than 350 types of dinosaurs presently known, none has been able to displace *Tyrannosaurus rex*, the king of tyrant lizards, as the leader of the dinosaur pack. Some species have been more agile, some more clever, faster, or more furious. But from the time that it was first described, in 1905, *T. rex* has been the supreme being of the Mesozoic.

Generations of devotees of horror films and books have long considered *Tyrannosaurus rex* their favorite villain. In a memorable scene from the classic monster film *King Kong*, the 40-foot ape battles with one of the dinosaurs that inhabit his fog-enshrouded island. His adversary, none other than *Tyrannosaurus rex*, was planning to make a meal of the heroine, the Beauty that Kong the Beast wished to have for his very own. Despite her obvious charms and desirability, the heroine would have been a mere hors d'oeuvre for *T. rex*, who lived in the midst of many gargantuan plant eaters and dined regularly on any creature it had a mind to.

Against a sound-effects background of the heroine's impressively relentless but ineffective screams, Kong comes to her rescue and kills the dinosaur by ripping its jaws apart. In real life the king of tyrant lizards would have had no difficulty making Kong his next meal. An adult *T. rex*, over 35 feet long, almost 20 feet high, and weighing seven tons, was the unchallenged superpredator of the late Cretaceous, 75 to 65 million years before the present (MYBP).

More recently, when Steven Spielberg chose *Tyrannosaurus rex* as an archvillain in *Jurassic Park*, he selected wisely. *T. rex* was doubtless one of the foremost hunters of the Cretaceous (not Jurassic) Period and was certainly the fastest as well as the largest carnivore that had surfaced thus far. The animal was built for hunting, killing, and eat-

ing live prey. So efficient, elegant, and formidable was its body design that nature engineered no improvement during the era of dinosaurs.

Tyrannosaurus rex was the absolute king of beasts during its 10-million-year reign. Its body was built to tackle and subdue the most powerful game around. For over a half-century after its discovery in 1902, scientists assumed *T. rex* to be a sluggish, lumbering giant that caught and fought its equally slow-motion prey in a quickened walk and was often reduced to scavenging. After all, modern oversized mammals cannot run fast, so how could an even more overgrown, underachieving reptile?

Studies of bones, muscle attachments, and footprints indicate that *T. rex* was anatomically correct for rapid movement and was probably quite agile. In fact, because of a formidable combination of size, speed, and weaponry, it was more than a match for anything that crossed its path. The muscles of its massive legs guaranteed high speed and endurance, punctuated by powerful kicks. *T. rex* was certainly the fastest animal for its size that we know of, and its birdlike limb design suggests that a speed comparable to that of a racehorse (30 to 45 miles per hour) was altogether possible. The rest of its body mass was directed to the head, with its stoutly constructed five-foot skull and killing-machine jaws powered by colossal muscles. It could open its traplike jaws a good three feet and, with gigantic serrated teeth three to seven inches long, polish off a cow-sized creature—horns, hoofs, and all—in three or four bites. In fact, its typical mouthful of meat would have been enough to feed a human family of four for over a week!

Recent research on an almost complete skeleton of *T. rex* suggests that it was warm-blooded. A warm-blooded animal the size of a tyrannosaur would require food almost constantly, and hunger would drive it to be a ruthless hunter. Not unlike the modern shrew, it would have to spend most of its time eating. Awed by the potential of this long-extinct beast, scientists have nicknamed it "Roadrunner from Hell."

Everything about *T. rex*'s body design spelled pursuit and kill. Its limbs tell the scientist that it could move much faster than most other predators. The two-clawed forelimbs were so small they almost appear underdeveloped or on their way to extinction, but most scientists no longer believe this to be true. Since the head of this huge carnivore was so massive, *T. rex*'s weight had to be reduced elsewhere on its front end. Its slender arms appear out of place on the colossal body, but they were far from useless. They were probably strong enough to lift 450 pounds, and the claws of the miniature limbs held the prey like grappling hooks while the jaws did their deadly work.

Several scientists still accept an early theory that *Tyrannosaurus rex* was a scavenger. The evidence indicates that at times *T. rex* would scavenge, just as a modern lion or tiger will do. When the food is just sitting there, why not? Fossil remains of a *T. rex* banquet were found near Alberta, Canada, where scientists have excavated at least 80 triceratops-like animals called *Centrosaurus*. About 70 million years ago a herd of these Ceratopsian dinosaurs attempted to cross a stream during a high flood. Several hundred of them did not make it and drowned in the raging waters. Their carcasses were washed downstream and piled up on large sandbars. This must have been an invitation for every neighborhood carnivore to a free lunch, and a dinosaurian feeding frenzy took place.

The scientists found many carnivore teeth among the bones, for the meat eaters of the day shed teeth virtually every time they fed. The teeth were replaced by new ones almost before the old ones were gone. Many of the teeth and tooth marks were those of tyrannosaurs, but most scientists now believe that scavenging was not an essential feeding style for the greatest land carnivore that ever lived.

T. rex was doubtless one of the foremost hunters of its day. Whether it stalked its prey or sprang from ambush, the end result was surely the same. One can almost imagine the huge beast springing from the forest and racing toward its chosen victim. What a sight to behold: a massive, menacing monster moving swiftly and with determination, its mouth open a full three feet and displaying innumerable teeth that glistened in the sunlight, while it emitted whatever kind of roar it made. It should have been enough to make the intended victim die from fright.

The tyrannosaur epitomizes the final greatest expression of the evolution of flesh-eating dinosaurs, terrible but elegant in form. In spite of being so efficient at killing and being the undisputed ruler of the Cretaceous world, *T. rex* fared no better than the dinosaurs on which it dined. With the close of the Cretaceous period 65 million years ago, tyrannosaurs, along with the rest of the dinosaurs, vanished from the face of the earth.

The classic picture of life in the primeval forests and plains of 70 MYBP is a battle between the world-class gigantic gladiators of the Cretaceous: *Tyrannosaurus rex* and *Triceratops*. Murals, books (including textbooks), and films have depicted this incident so often that it has been the automatic image for anyone who conjures up a vision of the life and times of dinosaurs. Although the triceratops would have been a most impressive adversary of *T. rex*, until recently there has never been any solid evidence that a tyrannosaur ever ate a triceratops.

Triceratops, among the most abundant of the Ceratopsian dinosaurs to inhabit the lands that would become North America, roamed the plains 75 to 65 MYBP in vast herds. They were enormous beasts that grew to lengths of 24 feet and weighed nearly nine tons. Their massive skulls, up to six feet long, weighed at least a ton and were adorned with a pair of four-foot, sharply pointed horns above the eyes and a smaller horn on the tip of the snout. Because the head was perfectly balanced on a pivot joint under the eyes, the neck muscles could toss it in any direction with great precision. When attacked, a triceratops had only to face an enemy, charge, and impale it. Surrounding its skull was a large protective bony shield that extended back over the neck. Broken triceratops horns and shields are common and, along with gouge marks, show that these ancient giants fought among themselves, most likely during mating season. When they collided over the affection of a waiting female, it must have resounded like a clap of thunder.

Armed with enormous jaw muscles, a heavy, parrotlike, beaked snout, and massive cheek teeth, *Triceratops* could have chewed its way through a tree trunk and probably did just that. And it may have also given the king some painful, appetite-quenching predinner abrasions. Generally these hoof-toed quadrupeds browsed quietly in herds on open plains, snipping bunches of tough plants with their pointed bony beaks and then chewing with scissorlike teeth.

Tracks of *Triceratops*, probably on migration, always show the young in the middle surrounded by adults. When attacked by a predator they probably formed a protective circle around the young. The sight of a ring of massive lancelike horns protruding from skull shields would have discouraged many a hungry carnivore. Recall a scene of medieval knights on horseback facing an enemy with weapons protruding in front of a wall of shining shields.

Many scientists do not doubt that the tyrannosaur would at times stalk the triceratops to within attacking distance and, with a loud roar, charge at top speed. Very likely the triceratops would turn and try to run away, leaving its flanks and back exposed to the tenacious teeth that could make short work of the exposed body. At times, however, instead of running, a huge adult triceratops would suddenly turn and face the oncoming monster, its four-foot lances poised directly at its adversary. With enormous calf muscles to power a quick charge, and knee-straightening muscles to provide great leverage, *Triceratops* possessed the agility necessary for a sudden, unerring attack. *T. rex* would not have time to slow up or to dodge the horns, as the momentum of its great body weight and speed of attack would carry it directly onto the threatening horns. The jungle would reverberate with roars of agony as the horns penetrated several feet into *T. rex*'s body. And the triceratops would walk away from the dying tyrannosaur, unharmed but with very bloody horns.

Until recently, scientists had no shred of evidence that such tyrannosaur–triceratops encounters ever happened. During the fall of 1992 such evidence was finally unearthed. Scientists from the University of California, Berkeley found a large adult triceratops hipbone that was 70 million years old. Several well-defined bites and marks showed where the carnivore's teeth dragged against the bone. To confirm the identity of the biter, the scientists filled the holes with dental putty; the resultant casts were identical with tyrannosaur teeth. Moreover, no other carnivores existing at that time were large enough to make such wounds.

Why was such evidence not found until after almost a hundred years of collecting fossil remains of these species? Considering the prerequisites for an animal to be preserved as a fossil, the chance of preservation for millions of years is incredibly rare. Compare, for example, the millions of bison that covered the plains of North America 150 years ago. Where are the billions of bones that would have been strewn over the area so recently? Occasionally a farmer may plow up an isolated bone, but most have weathered away naturally. Also, tyrannosaurs usually didn't leave much behind after they finished dining. Like modern hyenas, they consumed everything, including the bones. In the case cited above, *T. rex* must have been so full that it just couldn't swallow another bite!

Erotic Buzzards

One colossal headache for a utility company that supplies natural gas is the ever-recurring possibility of a gas leak in the widely dispersed pipe system. In the southwestern desert areas, the problem appears to have been solved in a most remarkable way. Often, making do with the materials and resources that occur naturally in the area offers the best solutions. In this case, the resource is the turkey buzzard.

In southern California and Arizona turkey buzzards are quite abundant and deserve their reputation as nature's cleanup crew. Any animal that dies in the desert will be located almost immediately by these birds, with their extraordinary sense of smell. The buzzards descend on the body and, in a relatively short time, strip the carcass of everything edible. When they leave, only bones remain for the desert sun to

reduce quickly to dust. Without the buzzard and other scavengers, the desert would become a disease-ridden habitat suitable only for bacteria to attack.

The buzzard's keen sense of smell would appear to offer very little help with gas leaks, especially since natural gas is odorless and tasteless. But strong-smelling substances are routinely added to natural gas because it is deadly dangerous and leaks must be detected readily. The simple solution: replace the usual added odor with one that the male turkey buzzard recognizes as the aroma of a female buzzard ready to mate. Their mating instincts instantly aroused, male turkey buzzards from miles around detect the leaks immediately and assemble in the area where they anticipate that some female will be soliciting their attentions.

Repair crews looking for a break in the line will search the sky for flocks of excited turkey buzzards. The birds can be found anxiously circling the precise location of the leak, jostling each other for the best position. Although their sharp eyes do not confirm the object of their ardor, their noses tell them that the invisible female is nearby. Excitement laced with frustration runs high, and overzealous buzzards have been known to make hostile advances toward the workers by flying over and bombing them, quite accurately, with droppings. Nevertheless, gas leaks are readily repaired, and the buzzards, when the scent of a female dissipates, vacate the premises and search elsewhere. Or they may become distracted by the aromatic, appetizing scent of a dead carcass.

A Medal of Honor

This is the story of a great military leader and a heroic unnamed elephant. Both were figures in the Second Punic War between Carthage and Rome. Three wars were waged by Rome to keep Carthage (a Phoenician settlement in Africa situated advantageously on the Mediterranean) from taking possession of any island or country that Rome might desire.

Most of the First Punic War (264–241 B.C.) was fought at sea, but toward the end of the conflict the Carthaginians used their secret weapon—an army of elephants—against the Roman legions. The Roman troops found themselves face to face with "Lucanian oxen," as the elephants were called. The beasts carried turrets on their backs filled with archers and, across the elephants' posteriors, broad shields to protect the brown-skinned warriors advancing behind.

The Carthaginian secret weapon failed miserably. Roman infantry and cavalry swarmed around the elephant squadrons commanded by General Hasdrubal and threw burning brands into their ranks. The frightened elephants began to trample the Carthaginian mercenaries instead of the Romans. In desperation, Hasdrubal ordered the stampeding elephants to be dispatched immediately by having sharp spikes driven into their foreheads. The riders, however, were too busy trying to jump safely from the elephants' backs to attempt to spike a stampeding, living tank. So Lucius Caecitius Metellus defeated Hasdrubal's elephant army at Panormus (now Palermo), Sicily, in 250 B.C. He captured all surviving elephants and led them triumphantly to Rome, where they were slain in the arena to the delight of screaming, bloodthirsty spectators.

Thirty-two years later, military elephants suffered the greatest single defeat of their entire combat experience. What caused their downfall was not the Romans but the malicious and ill-timed forces of nature. The occasion was the Second Punic War, which took place in 218–202 B.C. Hannibal (247–183 B.C.), the younger brother of Hasdrubal, was the Carthaginian general who led an army across the mountains and waged war against the Romans for 15 years on Italian soil. His tactics are studied in military academies, where Hannibal has been compared to such military leaders as Alexander the Great, Julius Caesar, and Napoleon.

Hannibal was the son of Hamilcar Barca, who had commanded Carthage's forces in Sicily during the First Punic War. After being defeated, the old general compelled the nine-year-old Hannibal to swear eternal enmity to Rome. The boy incorporated hatred of Rome into his rules for living; the idea became an enduring factor in his life. Hannibal's subsequent invasion of Italy was considered by the Romans to be the fulfillment of that vow.

Hannibal's plan was to conquer Italy from the north, despite African soldiers who were unaccustomed to cold and African elephants thoroughly inexperienced in travel through ice, snow, or mountains. With a courage born of recklessness and an army of 9,000 horsemen; 50,000 Carthaginians, Libyans, Iberians, and Nubian infantry; and at least 50 war elephants, Hannibal set out in May 218 B.C. from Nova Carthage (the modern city of Cartagena), and crossed the Pyrenees and Alps in an unparalleled forced march.

Hannibal had left his brother Hasdrubal behind in Spain to raise additional troops and to exploit the silver resources of the Iberian Peninsula to pay for the war. Hasdrubal and the new army were to follow later as reinforcements.

Crossing the mountains with an army of men and elephants was not without hazards. In the Pyrenees the Carthaginians were constantly harassed by hostile tribesmen who blended into the snow like ghosts, giving the oppressed men very little rest at night. But climate, weather, and terrain took the major toll of man and beast.

After Hannibal had crossed the Pyrenees, only 37 elephants remained. But in the Alpine passes, the battle with the elements turned into sheer tragedy, and by then there was no turning back. The peaks were covered with snow, and whatever paths there might be were frozen over. The mercenaries and elephants, weak from hunger and forever cold, constantly slipped and fell into yawning crevasses. Some of the soldiers, frostbitten and too weak to walk, were able to ride on the elephants. When a beast slipped over the edge, its frightened trumpeting blended with the human cries as they fell together into the chasm. Doubtless the remains of many of these casualties of war still lie frozen in remote parts of the Alps, yet to be discovered.

When Hannibal finally reached the valley of the Po, his army had shrunk to 20,000 infantry, 6,000 cavalry, and eight elephants. With this mere remnant of a once mighty force, Hannibal had to face several

Roman armies, each many times the size of his. Fortunately his army was boosted to 40,000 by the immediate addition of Gauls, the traditional enemies of Rome with a nurturing hatred of Romans that equaled Hannibal's.

The war continued in Italy for years, and with fewer and less skilled troops Hannibal did not suffer a single defeat. In 209 B.C. Hasdrubal crossed the Alps with a large army and moved south to join his brother Hannibal. Knowing that the two armies combined could move against Rome, the crafty Romans engaged Hasdrubal in battle and won. Hannibal refused to believe the news, but he quickly became a believer when Roman horsemen galloped through his camp and tossed a head at his feet. With beard bloodied and glassy eyes staring, the head was not difficult for Hannibal to recognize as his brother's.

Almost at Rome's back door, Hannibal may still have been able to conquer it. But because Scipio Africanus was invading North Africa, Hannibal was called home to defend Carthage. In 202 B.C. Hannibal's new army met Scipio's outnumbering troops and cavalry at Zama, North Africa. He was defeated and, with his army destroyed and Carthage stripped of its navy and overseas possessions, the valiant warrior escaped into private life. He eventually accepted exile rather than continued Roman threats. In 183 B.C., no longer able to live with his disgrace, Hannibal took his own life to be free of further humiliation.

The tragedy of lost wars does not end here, because one elephant did survive the entire campaign. Fifteen years after setting out from Carthage and having traversed Spain, Gaul, and Italy, crossed the Pyrenees and the Alps, taken part in at least six major battles, and come within three miles of Rome, this lone surviving elephant returned safely to one of the great stables of Carthage. Of all animals that have participated in and aided man in military adventures, this truly heroic beast would have had the best claim to a medal. In the United States he could have qualified for the Congressional Medal of Honor.

Not even his name remains. In fact, he may not have had one. But as a lone elephant, he could have been called "The Lucanian Ox."

How Not to Find a Volcano

During the late 1980s a rare underwater volcano in the South Pacific erupted only 130 feet beneath the *Melville*, a San Diego–based research ship. Erupting from the MacDonald Seamount, located 650 miles southeast of Tahiti, the volcano sent gas-infused rocks clattering and clanging against the steel hull of the *Melville*. The eruption transformed the blue-green water into a churning, boiling, dark brown muck and enveloped the vessel in a swirling tumult of murky water, gas bubbles, and hot volcanic rocks. The boat rocked and rang with "horrendous clangs and clamors" but suffered minimal damage. The chief scientist radioed back to his home base, "Large gas and steam bubbles burst at the surface with chocolate-colored water and, best of all, steaming lava balls too hot to hold in bare hands."

The eruption coincided with the scientists' arrival to study volcanic activity in the general area. They saw the event as an extraordinary bit of luck—being at the right place at the right time. One team member related, "I think it's fair to say that nobody has ever seen anything like this before."

The man was evidently unaware of a similar event that had occurred years earlier. The eruption under the *Melville* was not particularly destructive; the scientists were lucky. In 1952 a marine volcano erupted violently about 25 miles south of Tokyo. Huge fountains of steam, water, incandescent gas, and ash shot high above the surface of the sea. As soon as the eruption appeared to have run its course, teams of scientists in several ships set out to study it. When the ships arrived at the eruption site, they spread out trying to locate it. One of the ships discovered its precise position because suddenly, without fanfare, the volcano erupted very violently directly underneath it. The personnel on the other ships watched in stunned horror as the ship, sailing peacefully over a calm bit of water, was abruptly blown to pieces by the undersea eruption. The ship was completely destroyed, along with everyone aboard.

Fish Out of Water

To keep their eggs safe from snapping jaws, drifting into unsuitable waters, or any of the other fates that may befall small creatures in the sea, many fish have devised unusual tricks. They may keep the eggs in their mouths, bury them, build many sorts of nests, cement them to some stationary or floating object, or deposit them in a pouch appropriately provided by the male.

Fish that have no such safety net for their fry must lay a sizable number of eggs. This is the case with fish that range the seas and lay pelagic (floating) eggs. The female cod and turbot, for example, lay between two million and nine million eggs at a single spawning. The oceans are not overrun with cod and turbot because, by the time the tiny fish are two inches long, fewer than a dozen will have survived from each million eggs: one survivor out of 83,000 eggs. Most of the rest become dinner for the many creatures of the sea.

Among the most remarkable fishes are those that, knowing the dangers of the waters, lay their eggs on land.

Miniature Sprinkling Systems!

One fish that lays eggs on land is a species of characin, *Copeina arnoldi*, whose family includes such aquarium favorites as the tetras, x-ray fish, and pencil fish and such sinister aquarium rejects as the piranha. *C. arnoldi* is a three-inch blue-gray fish that lives in slow-moving South American rivers.

This little fish is spectacular only because it lays its eggs out of the water. The male steers a willing female to a spot where the leaves of a hanging branch are just an inch or two above the water. The pair then hook their fins together, leap out of the water, and cling to a leaf for about 10 seconds. During this brief period the female lays a dozen or so eggs in a jelly mass that adheres to the leaf. The pair return to the water and repeat this maneuver until all the eggs are deposited and then fertilized by the male. The spawning completed, they unlock fins, and the female swims off in the never-ending search for food. The male, as a fish version of a househusband, is stuck with minding the nest.

Laid out of water, the eggs are safe from menacing jaws but are in danger of being dried up by the sun. Keeping the eggs moist is a job for the male *C. arnoldi*. He returns to the egg mass every quarter of an hour and splashes water on the brood with his tail. If he fails to return, the nest will dry up. The only thing that will sabotage his standing appointments with the offspring is an invitation to dinner by a larger fish—he being the dinner. But by and large dad returns faithfully as scheduled, joyfully splashing water on the eggs to keep them healthy and wet. Fortunately for him they hatch in three days (288 trips, in case he is keeping score). The fry then drop into the water and swim away without so much as a nod of appreciation for their exhausted father.

"Dancing Fish"

At certain times of the year around midnight, under a full moon, for several hundred miles along the California coast, the sea virtually explodes with thousands of silvery fish that appear to perform a mysterious dance on the beach. They sparkle like diamonds in the moonlight. When this spectacle is at its highest intensity, so many fish may be visible on the sand that they create the illusion of a silver canopy covering the beach.

The fish's weird antics give the impression that they are dancing, from which grew the Native American legend of the Dancing Fish. But the dance is actually their unusual spawning ritual, for these are the grunion.

Because grunion eggs will die if kept immersed in water, an alternative, nonwatery haven must be found for them. Therefore the fish have selected a 100-mile strip of beach, from Point Conception, California, to Baja California, Mexico, for their haunting natural drama. These six- to seven-inch fish, *Leuresthes tenuis*, have a long breeding season, from late February until early September. But actual breeding can take place only on the three or four nights that follow the highest tides during a new or full moon.

As the time to spawn nears, the grunion stops growing. Instead, its energy is channeled into reproduction, and its sexual organs grow larger. Within the female the first eggs begin to ripen. Yearling females produce only about 1,000 eggs per spawning, but the number is tripled in older fish. Since a single female can spawn up to eight times a season, many thousands of eggs can be left ashore by one fish.

With incredibly precise timing and a knowledge of tides that the U.S. Coast and Geodetic Survey might envy, the female prepares for her trip ashore. She must lay her eggs in the sand after the peak tide or they will be washed away by higher tides, and the eggs must be protected for at least 10 days while they are incubating.

The grunion gather offshore for an hour or more before the run. When the time and tide are right, the grunion allow themselves to be flung high on the beach by an incoming wave. The sand is immediately carpeted with quivering silvery bodies twisting, squirming, shimmying, or (as legend would have it) frolicking in a frenzied dance.

The fish strive mightily to reach a point on the beach just below the high-water mark. Stranded high and dry, unable to breathe, the females nevertheless arch their bodies vertically and burrow violently into the sand with their tails until only their heads and pectoral fins are clear. This takes about 10 seconds. Several males writhe around each female and discharge their milt (sperm-containing fluid), which trickles down to fertilize the eggs deposited in the sand. The males will wriggle back to the safety of the sea, but the female will rest for a moment, making a tiny, high-pitched grunt. (This behavior gave the species its name, derived from the Spanish word *grunion*, for "one who grunts.") She will shortly follow the males back into the sea. Depositing and fertilizing a cluster of 1,000 to 3,000 eggs is a 30-second task—with arrival and departure time, one minute.

The water of the tides for the days following doesn't reach the buried eggs but does add more layers of sand. This gives the eggs extra protection from sun, storm, and probing beaks of predatory shorebirds.

The eggs are ready to hatch in about 10 days, but nothing happens until the next flood tide about a fortnight later. Then incoming waters swirl the sand around the eggs, which hatch with explosive suddenness. The young grunion shoot into the water on the wave that uncovered them. And there they remain, not far from shore, at depths of no more than 60 feet, until spring tide of the next year, when it will be their turn to become a dancing fish in the light of the moon.

How do these remarkable fish know when and where to spawn, and what triggers the spawning runs? What activates the extremely accurate internal clock that makes them beach themselves with such precise timing? Do they sense variations in the force of gravity that produces the high tides, or changes in water pressure? Or do they somehow respond to changes in the intensity of the moon's light? When the survival of their young hangs in the balance, animal species instinctively know whatever they need to know in order to survive. Those that do not will face extinction!

The Real Middle of Nowhere

People often describe unfamiliar, uninhabited places, such as a remote section of the Mohave Desert, as being in the middle of nowhere. This phrase would seem to fit a large array of locales, anywhere on the planet, so long as the spot is barren, isolated, far from human habitation, and uniform and uninterrupted from horizon to horizon. It could be a desert, a jungle, a mountaintop, a wheat field, or even an ice floe.

There really is a "middle of nowhere." Scientists have identified a great expanse of nothing but water as the most remote spot in the world. This spot in the South Pacific at 48 degrees, 30 minutes south, and 120 degrees, 30 minutes west, 1,660 miles from the nearest land, is Pitcairn Island. The mutineers of the HMS *Bounty* were lucky to have stumbled onto that two-square-mile piece of land, since collectively a total of 8,657,000 square miles of uninterrupted water surround it.

The Golden Age of Medical Quackery

The story of medical quackery probably began shortly after the first cry of pain resounded through a primitive jungle and brought the first physician. Before long the healer realized he could be more effective, and the afflicted one more responsive, if the treatment were laced with a touch of hocus-pocus. In Medieval Europe *quacksalver* was the term that identified a person who sold salves by noisily hawking, or "quacking," the wares. Eventually *quack* became the word, because to some the quacksalver sounded like a noisy duck. The charlatans who were the early quacks were often, then as now, simply dishonest folk out to make a quick buck by taking advantage of the gullible, the fearful, and the ignorant.

First exposed by Rhazes, the wise man of medicine of ninth-century Persia, medical quackery was well recorded during historic times. Europe and Asia had long been rife with bold tricksters, incredibly successful at fooling the public. Rhazes enumerated a number of their stratagems. Since almost any complaint could be blamed on the presence of a foreign body, the pretender "removed" it. Palming the real or fabricated foreign body and making an incision where the patient could not observe it (the ear, nose, mouth, and back of the head were ideal), he drew blood. He could then show the patient the evidence: small rocks, lizards, worms, frogs (usually artificial animals composed of liver). The patient, much relieved that the cause of the affliction had been removed, expected to feel much better as soon as the surgical wound healed.

Eighteenth-century England is remembered as the Golden Age of Quackery, since Queen Anne patronized and gave credibility to myriad swindlers and frauds. One of the more successful of the self-confident but unqualified healers was a Frenchman, Jacques Beaulieu, who finally became licensed as a stone cutter, or lithotomist (one who removes kidney and bladder stones). He traveled through Europe removing gall stones and kidney stones, often successfully. He wore the habit and broad-brimmed hat of a Franciscan friar to guarantee security and free hospitality. He also gained wealth (which he left to charity) and immortality, for he became known in song and rhyme as Frère Jacques (Brother John).

America's first renowned quack was Dr. Elisha Perkins, who in 1796 patented a device guaranteed to cure anything by simply moving "Perkins's Patented Metallic Tractor" downward over the ailing part of the body. By moving it slowly and carefully, the doctor could yank out any disease known to humanity. Perkins's device was the rage of the time, and he gained considerable wealth and respectability. A notable who used Perkins's appliance was George Washington, whose entire family found it therapeutic and said treatments gave them a sense of well-being. Even Perkins became convinced that this device actually did work, although its special properties were among the natural wonders that he didn't quite understand.

The purveyors of quack medicines surfaced early in colonial America. Capitalizing on the ordinary person's ignorance and fear of illness, pain, and death, they needed only to convince ailing people that they could relieve suffering. Vendors of herbs, pills, potions, patent medicines, or their own concoctions and lay practitioners of indiscriminate medicine made the rounds of country villages. Indian potions were popular and trusted cure-alls. After all, the Native Americans were in tune with nature, weren't they? Hadn't their knowledge of roots, barks, herbs, gums, and leaves—to say nothing of buffalo tallow—been keeping them strong and well for generations? The mountebank, half showman and half quack physician, would tour the country with a wagonload of wares and material for a spectacular show. Even before the Revolution, enough of them were around that several colonies passed legislation to "suppress and control mountebanks dealing out physick and medicine of unknown composition."

During the 19th century, medical quackery really flourished in rural America. Driving wagons profusely decorated with ads proclaiming the effectiveness of their potions, medicine show entertainers toured the country hawking their wares. Despite disapproval from town fathers and the pulpit, the mountebank delighted crowds of entertainment-starved people. The sideshow kept the crowds coming and set them up for a proper shearing.

Often the only medicine sold was a cure-all. By cleaning impure blood or regulating the action of the stomach, liver, kidneys, lungs, or some other organ, this "wonder medicine" would cure malaria, chills, biliousness, nervousness, rheumatism, headache, catarrh, heart palpitations, wakefulness, piles, dyspepsia, skin disease, pimples, and dozens of other afflictions. And the clever huckster could extend the potential of the elixir to accommodate any distress suffered by an audience member, be it baldness, boils, or a broken arm.

Most of the medicines were priced at one dollar, a significant amount in the 1800s. But the seller might display sale prices to tempt a reluctant crowd or reward a responsive one. Usually there was a money-back guarantee, but by the time the still-ailing patient was ready for a refund, the doctor would have long since changed his location, appearance, and name.

Many early medications that guaranteed a euphoric feeling while curing myriad ailments contained cocaine. One French tonic wine combined two ounces of fresh coca leaves per pint of Bordeaux wine. This potent concoction was praised as an elixir of life by such notables as Jules Verne, Emile Zola, Sarah Bernhardt, Thomas Edison, and John Philip Sousa.

Professional quacks may have been amazed at their own occasional success. They should have recognized, however, as have poets and philosophers, that time, patience, a good laugh, and a long sleep are the best cures. That patients regained health from some quack machine or medication should have come as no surprise; doctors have long known that 60 percent of the time the body, untreated, will cure itself.

One entrepreneurial genius in quackery (low overhead, high profit, great demand, quick results) sold diet pills in the 1930s. They were guaranteed to make the user lose weight rapidly, and they certainly did work but with very undesirable side effects, including several deaths. Chemical analysis of the pills showed that they were nothing but tapeworm eggs. Unfortunately the woman who sold them disappeared without a trace.

Peddlers of wizard oil and other cure-alls continued to ply their wares long after the Food and Drug Act of 1906 began to cramp their style. One of the last was a state senator from Louisiana who created Hadacol, the universal potion of the South. By adding honey to B vitamin tonic and lacing it generously with 12 percent alcohol, he could almost guarantee a gallon jug of his quick picker-upper in a convenient corner of every southern home. Stupendous advertising, high-pressure selling, and Hadacol Caravan shows (with a smattering of show business celebrities) got out the word. Money could not buy admission to the spectacular shows—only Hadacol boxtops could (two for an adult, one for a child). The medicine show had definitely reached new heights. Then, in 1951, the boom began to bust, and by 1954 free shows to sell products had been taken over by television on a scale far beyond anything the medicine show huckster could have ever dreamed of.

During the uranium boom of the 1950s, quackery again reared its crooked head. Near many cities and towns in the uranium-producing states such as Utah, New Mexico, and Colorado, enterprising souls set up uranium sand houses. These crudely built, somewhat circular structures had a single large room with a sand-covered floor. Patients could sit on benches along the walls, but most sufferers preferred to bury themselves, or at least their ailing body parts, in the radioactive sands. An hour or two of absorbing the radioactive rays was considered a therapeutic treatment, and the "doctor"-proprietor usually recommended more than one treatment. The patient had a relaxing, stress-free rest, along with a gentle fleecing, and left much better able to face the world.

The clean, inviting beach sand really was radioactive. The owner would obtain a radioactive mineral such as carnotite, crush it to a powder, and sprinkle it sparingly throughout the sand. Cautious use of the mineral had nothing to do with fear that the uranium might be harmful but was merely a cost-cutting strategy. A geologist in Utah checked out the radioactivity of a uranium house with his Geiger counter. He found it to be weakly radioactive, barely above the level of residual or background radioactivity. Luckily, extravagance was not one of the owner's shortcomings; no one ever used enough uranium to cause leukemia or any radiation-related ailments.

As with the quest for the universal solvent, the search for a panacea to cure all ailments has not ceased. One grandmother sent for a product, advertised by a doctor in an Atlanta, Georgia, newspaper, that was guaranteed to immunize her from all afflictions. What she received was a gold pencil-shaped instrument that was supposed to contain radioactive material. The instructions were to put the pencil in a glass of water and leave it there overnight. Then she was to remove the pencil from the glass and drink the water. Despite the hoax, the grandmother lived to the ripe age of 97. The amount of radioactive material in the pencil was so slight that she was not harmed by it, and she did receive the advantage of an extra glass of water each day for about 15 years.

In the past, and currently as well, old wives' tales have been the source of much medicine, some illogical quackery and some effective treatment. A Palm Springs, California, resident recalls how, throughout his early childhood, his mother would wrap slices of raw onion in a cloth that she tied around his neck whenever he complained of a sore throat. The only noticeable result was that the onions would disappear. This happened quite regularly, because complaints of a sore throat increased in frequency as the boy acquired a fondness for onions. His partiality to onions has remained with him, and 50 years later he cannot recall ever having had a cold last more than two days, even when the cold virus raised havoc with his family.

Onions do have therapeutic value and have been recommended medicinally over the centuries for dog bites, insomnia, pneumonia, diabetes, overweight (only 34 calories), rheumatism, baldness, and gunshot wounds (apply topically). To this list our onion enthusiast would add cold prevention. He does admit, however, he must always stay downwind from his friends after indulging in his favorite aromatic vegetable. Perhaps keeping a distance from friends and family is the real reason that he has remained relatively free of colds!

Lost: One Ungrateful Slave

About 3,000 years ago an Egyptian gentleman was quite upset that one of his favorite slaves had run away. Why should the slave do such a thing after all the owner had done for him? The slave had been whipped only when he deserved it, and he was fed as well as the household dogs. With all these acts of kindness, why should the man want to be free? The only logical conclusion was that the runaway slave was an ingrate.

The distressed nobleman insisted that the slave be brought back and punished. So, on a bit of papyrus, he offered a reward for the return of this ungrateful runaway slave.

Whether the slave was returned or not will never be known, since no record exists of anyone's collecting the reward. Anyway, this is of little historical importance—except, of course, to the slave. What *is* of consequence to scientists and scholars is that this bit of papyrus, preserved in the British Museum, remains the oldest known piece of advertising copy.

The Unsuicidal Lemming

In 1967, on a balmy spring day in Alaska, millions of burrowing hamsterlike rodents emerged from the ground. No one walking in the area could avoid stepping on the ground cover of squeaking lemmings. As if seized by mass madness, the animals assembled into a horde stretching for miles and rushed straight across the Alaskan tundra for nearly 125 miles. At a steep cliff near Point Barrow, the endless ranks plunged into the cold Arctic Ocean. The frigid water became the grave for the millions of lemmings that completed the trek. Witnesses insisted they were committing suicide. And so it would appear.

These rushes into the sea have provided the occasion for many lemming legends, particularly in Norway. The lemmings, the Norwegians maintain, are obeying an instinct to migrate to sunken Atlantis, or even to an ancient Greenland supposedly connected to Scandinavia in the geologic past. In earlier times people believed that lemmings fell from the sky like rain. How else could one explain the fact that suddenly the land was covered with hysterical, squeaking lemmings scurrying to nowhere?

Attempts have been made to predict the "lemming year," the cycle of population surges, as every seven, five, or three years. But the irregular explosions seem to follow several years of ideal weather conditions that produce increased vegetation and, therefore, lemmings aplenty.

Normally a small, inconspicuous animal, the lemming is a common rodent throughout the northern latitudes of all continents. Four to six inches long, with an inch or less of tail, thick, water-repellent coat, blunt muzzle, small eyes, and ears hidden under fur, it is well built for resisting cold.

In the summer, lemmings feed well on the abundant mosses, lichens, and other arctic vegetation. They store additional leaves and roots in holes and crevices between rocks. Rather than hibernating in winter, they dig shallow burrows in the ground or, if the earth is frozen, just under the snow. In the fall a long digging claw grows out from under the normal claws of each foot so they can cope with the digging demands of a frozen landscape. They live off roots, stems, and underground moss along with the food stored during the summer.

Large colonies of lemmings live together, congested but snug, and safe from foxes, lynxes, weasels, wolves, owls, ravens, hawks, and other predators. Breeding continues almost uninterrupted, and the winter season produces several litters. In spring digging claws are shed, and the lemmings leave their snow burrows. The complicated maze of

well-trodden snow and ice that formed their tunnels is often the last thing to melt, leaving a lacy network of ice as a spring decoration.

Lemmings are prodigious producers. The average lemming litter has five or six young, and five or more litters may be born to a single female in one year. Although most mothers deliver their first young at age four to six months, they have been recorded as giving birth when 39 days old. Gestation taking about three weeks, the precocious mother would have become pregnant at the age of 20 days. With no reproductive controls and with several females from the first and second litters adding their families to the heap, the population grows explosively. And explode it does!

By late spring the passages of the underground burrows have become choked with lemmings. Roofs and walls cave in when their city can no longer accommodate the swarming crowds. Clamoring for space, the normally docile lemmings become aggressive toward each other; fights break out incessantly, and bickering and biting are everyday occurrences. Finally each animal obeys the impulse to get away, and "refugees" emerge from every burrow. Within a few hours they have eaten all the vegetation in their area and begin to mill around in wild confusion, gripped by panic. They run and leap in all directions, squeaking, barking, and gnashing teeth.

Eventually one lemming will cross a frontier that had been the outer limit of their world. Others follow until millions congregate into an enormous throng. They advance arbitrarily in all directions. They climb down mountains, only to meet another procession of lemmings climbing up. By some grim paradox, each lemming, at the onset of the migration, is obeying an unconscious instinct to flee from its perilously increasing numbers, but the migrating masses unwittingly take their problem with them.

During the march, the lemming becomes aggressive out of all proportion to its size, even to humans it may encounter. It will square off at a passerby, uttering doglike barks, gnashing its teeth, and biting. The stubborn little rodent holds its ground against any adversary, snarling and biting at a fox or weasel intent on making it the next meal. At times so many lemmings are killed by automobiles and railroad trains that massive removal of bodies by the National Guard becomes necessary for basic sanitation.

Individual hordes, sometimes over a hundred miles long, tend to move in a straight line toward some point on the horizon. Passing through villages, they run through rather than around any structure that is accessible. Human occupants must watch helplessly from a rooftop until the throng has passed and even more helplessly as it ravages newly planted crops. Those aware that this is a "lemming year" delay planting until after the caravan has moved on. Lakes and rivers are no problem; the lemmings, excellent swimmers, paddle across. Should a small boat happen in their path, they will climb aboard on one side and jump overboard on the other while the human occupants try to remain inconspicuous.

For almost all of the remaining lemmings, the expedition is over when they perish from stress, exhaustion, hunger, or predation. They have inadvertently provided a bounteous feast for all carnivores, including birds of prey and even fish during their march. The number of lemming marchers has been reduced many thousandfold by the time the obsessed travelers reach the seacoast. They plunge into the surf, swimming in dense squadrons farther and farther out toward that spot on the horizon. Mariners miles from land have reported that their ships appear to have run aground on what looks like a solid furry landmass of lemmings that extends for miles. The little rodents, kicking and paddling, swim continuously toward some vague paradise. Eventually they reach eternity.

The lemmings, of course, are not trying to commit suicide. They are merely searching for food and living space and carry their search to its inevitable conclusion.

There are exceptions. Some lemmings, observing the mass flights from their native habitat, recognize that the situation at home is correcting itself, and they stay right where they are. Other restless migrants occasionally have the good fortune to come upon territory suitable for their needs. For these lucky few the migration is over. They are the dropouts who can begin a new life.

Massive population explosions that drive most of a species into a frenzied march toward death seem to serve no purpose. But the lemmings that remain from one explosion seem to be stronger, calmer, and better able to resist the migratory impulse. They are the new generation that will prosper when the plague of overpopulation has subsided. This appears to be a graphic example of natural selection and the survival of the fittest. Also, the lemmings can provide the human race with the merciless example of the catastrophic madness to which an exploding population can be driven. Somehow the resemblance to the teeming human multitudes scuttling aimlessly about the world's cities in mass confusions and self-destruction is not easy to ignore.

Fossil Forests of Yellowstone

Of all the wonders in Yellowstone National Park, one of the most fascinating is usually passed over by tourists. In the Amethyst Cliff area of the park, exposed to view on the steep slopes, is a 2000-foot-thick section of layered rock that records the birth and death of 18 successive forests.

During the ancient Oligocene epoch, in the area now known as Yellowstone National Park, numerous explosive volcanoes caused huge amounts of ash to shower down on the land, burying large forested areas. Instances of ash "showers" were interrupted by periods of dormancy, often long enough for the forest to regenerate. So a new forest would grow and thrive on top of the layers of ash that buried earlier forests. Rock layers were deposited one on top of the other. One layer devoid of any remnants of plant life would be overlain by the remains of a younger forest destroyed by a volcanic shower of ash.

What happened in prehistoric Wyoming is written very concisely in the rocks. In the million or so years of time recorded there, huge forests developed, only to be disrupted when the area's dormant volcanoes came to life. Daylight was turned into sudden night, the air choked by black clouds of eruptive dust crisscrossed by violent lightning bolts. When the ash settled out of the atmosphere, where beautiful forests had once stood was now a sea of dead, burned trees, most blown over by the volcanic blasts. They resembled huge matchsticks strewn in the direction of the blast; nothing appeared to be alive.

As in all major disasters, however, some life always manages to survive. Eventually, after the blackened sky cleared and with hot volcanic ash a thing of the past, survivors began to stir. In relatively short periods of time new buds pushed through the soil covering, and a new forest was born. This rebirth, repeated over and over again during at least a million years of geologic time, is pictured in Amethyst Cliff.

Faulting has uplifted the areas of fossil trees and, aided by recent erosion, has laid bare the sides of several hills, exposing the successive forests, one on top of another. Doubtless several layers of buried forests still lie in the subsurface, not yet exposed by erosion. Many of the trees exposed to view on the sides of the hills are truly unique, for they stand upright in the same position in which they lived and died—over 40 million years ago.

Formula for Survival

Scorpions are probably front-runners among the creatures that nobody loves. Every one of the 700 known species, ranging from one-half inch to over nine inches in length, is characterized by a poisonous stinger at the end of the tail. Size is not an exact formula for the power of the poison, and awareness of the smaller scorpions is always prudent. (Few people need to be reminded to stay clear of the seven- to nine-inch models!)

The formidable reputation of scorpions can be traced back to their ancestry in early Paleozoic time. By the Silurian period, 430 MYBP, their ancestors, the eurypterids, were well-developed sea dwellers. Commonly equipped with brutal claws, spines, a vicious stinger, and a generous poison gland (the first for any species), eurypterids were among the most powerful rulers of the crowded Silurian seas. There were many varieties, typically 5 to 30 inches in length, but some giants attained a length of nine feet. These creatures gratifyingly became extinct, and unknown descendants crawled out of the densely populated seas onto the driest of land. Here on the deserts of the earth their evolutionary nieces and nephews, the modern scorpions, have been at home for millions of years—the first poisonous creatures to inhabit dry land.

In the early days of civilization people in warm, dry climates such as in the Near East and Egypt walked around in open sandals. This was particularly hazardous because several deadly varieties of scorpions abound in such climates. Open sandals made the people easy victims, but for many generations they seemed to accept it as an unavoidable natural hazard. Therefore, many died from scorpions stings. One group of disgruntled people finally decided to take steps to protect themselves from scorpions. They were the ancient Assyrians, who simply invented the boot! For this invention the world is eternally grateful, as long as we remember to beware of the old scorpion-in-the-boot trick.

Similarly the scorpions helped to develop the art of carpet making, for they were among the creepy crawlies that made sitting on the grass or sleeping out of doors hazardous to health and well-being. The Moslems, required to prostrate themselves regularly for prayers, were determined to make devotion to the Supreme Being less lethal. A portable "floor" of sheep's wool, laced naturally with lanolin, discouraged invasion from all kinds of arthropods, which usually avoid greasy substances.

In the tropical and subtropical regions of the world, dense populations of scorpions live in a number of specific areas. They concentrate in and around human dwellings, generally under anything, ready to crawl out at night. The assortment of species in each region, including the southwestern United States, ranges from harmless to painful to deadly. Many barefooted citizens in warm regions are victims of the scorpion, and in a number of places, including the United States and Mexico, scorpion stings cause more deaths than do snakebites.

The scorpion population was so high during the late 1940s in Durango, Mexico, that the government offered a bounty for them. In three months over 100,000 specimens were turned in for bounty, and this harvest was repeated several times. The bounty is no longer in effect, and Durango remains as overrun with scorpions as ever. On the other side of the world, researchers in an infested area in the Namib Desert and others in Bombay have encountered scorpions every few feet throughout the countryside. One scientist collected over 14,000 specimens in a single night.

Scorpions unfortunately do not avoid human companionship and all too frequently do not hesitate to share a bed or sleeping bag with any accessible human. The sting can be fatal to a child or to an infirm or elderly person. The scorpion's reputation is tarnished further by this record, although its basic problem is simply that it is overabundant around human habitations. Since stinging humans does nothing to keep it alive and well, the scorpion would doubtless avoid people if it could. Humans who live where scorpions are common would do well to be careful and observant at all times and avoid the scorpion.

The problem of keeping visible has been partially solved by several species of desert scorpions. They fluoresce under ultraviolet light and are brightly visible when a black light shines on them. This happens often in fluorescent mineral country, where prospectors for such hard-to-distinguish minerals as wolframite, fluorite, calcite, and scheelite often locate the minerals with a black light.

To keep visible light at a minimum, mineral hunters explore the desert at night. Almost all creatures are stirring because the temperatures after sunset are more tolerable. Many a prospector, reaching for an object glowing in the dark under a black light, has received an agonizing sting instead of the thrill of discovering sought-after minerals. They quickly and painfully learn not to grab every object that glows in the black light.

The scorpion is a solitary, nocturnal arachnid. Its principal food is other arachnids, particularly spiders, or even other scorpions, since they are not immune to their own venom. An excellent hunter, the scorpion can eat until it almost bursts when food is plentiful. It converts extra food into carbohydrates and when sustenance is unavailable reduces body metabolism and consumes the stored carbohydrates. By living off the "fat of the abdomen," the scorpion can sustain itself for a full year without a morsel of food. Truly a formula for survival.

Although scorpions generally keep to themselves, the situation changes dramatically at mating time. Because each partner could regard the other as a meal rather than a mate, many protective rituals are part of the courting procedure. They begin their relationship by locking pincers and intertwining tails to immobilize their sting apparatuses. They gyrate backward and forward in a strange primitive dance, and after some time their shuffling steps will have cleared the dance floor of much of its debris. Then the male extrudes a packet of sperm on the cleared area and heaves the female directly over it. The mating is then completed. Unless the male unlocks and runs, a wedding feast takes place attended only by the female, for she will sting the male to death and consume him. This cannibalistic ritual, performed by a number of spiders and insects, is a most expedient way to provide an excellent source of protein for the growing brood of eggs.

Development of the eggs can take several months to a year, depending on the species. Several score are born alive, ready to climb aboard their mother. They often cover her entire dorsal area. Consuming the food stored in their bodies, they remain on the mother's back for a week or two until they molt and can fend for themselves. This marks the end of maternal responsibility. Neither offspring nor mother will have any further contact with the other, unless the mother and a male offspring meet by chance during another mating season. All of this is part of an intricate formula for survival.

A few unconventional people keep scorpions as docile, well-trained pets. Itinerant performers in West Africa capture and tame large scorpions, allowing the arachnids to live inside their voluminous clothing. The poisonous tail spine is rarely removed, yet these potentially dangerous creatures wander freely over their owners. They are truly well-tutored pets and will respond to commands given to them by a mere touch of their owner's finger. Strangely, scorpion training has not been included among the top 10 high-stress occupations, nor is it high on the list of career objectives.

The Song of Spring

In parts of the world where season changes are conspicuous and dramatic, people often rhapsodize during the cold winter over the coming of spring. They describe spring as a time when the crocuses and daffodils burst into bloom, accompanied by songs of birds. The robin especially has become the symbol of spring, since it arrives in upper latitudes as soon as the snow melts and certainly does its share of singing.

For people who live where the robin spends the warm seasons, spring officially starts when they hear the first notes of the robin's song. The bird seems to be celebrating the new season. Scientists know, however, that its singing has another purpose.

The robin's song proclaims to other birds, including other robins, that it and its mate have laid claim to a certain area to raise their offspring. In other words, the song of the belligerent, bad-tempered robin is really a warning to other birds to back off. Far from being a celebration of the arrival of spring, this song is a robin's war cry. When a rival bird intrudes on robin territory, the song becomes louder. If the rival bird doesn't retreat, the robin furiously attacks—still singing!

Canine Sky Diver

As the World War II soldier stood in the door of the C-47 waiting for the quick signal to jump, visions of his chute not opening were not unusual. But after a stern tap on his boot, he stepped out of the door and into space. He could feel his body tumbling and the sudden jerk as the chute opened. If this was a combat jump his troubles were just beginning.

During World War I, for combat pilots, wearing a parachute was considered cowardly, and so they didn't. Many a death could have been prevented but for this misplaced medieval emphasis on pride in noble bravery. The situation was quite different during World War II, when any member of the service permitted to board a military aircraft had to be equipped with a parachute. Today people skydive for pleasure, and those who conduct the jumps make a good living giving clients a jolt. Earlier generations of paratroopers, who were at times helped out of the plane by a foot to the back, have difficulty imagining parachuting as a safe, thrilling sport.

Parachuting was invented more than a hundred years before the Wright brothers conducted their first flight at Kitty Hawk, North Carolina, in 1903. (A parachute would have done them no good, since they didn't go high enough.) The inventor of the parachute was a French balloonist named Jean-Pierre Blanchard. Mr. Blanchard was playing it safe, however, for he did not make the first jump. His dog did!

In 1785 the inventor of the parachute secured his dog inside a basket and tied down the lid. Ascending to a relatively great height in his large balloon and with gusto, the courageous Mr. Blanchard threw the basket with the parachute attached overboard. The wind immediately filled the primitive wind-catching device, and it opened. The dog was unhurt in the somewhat rough landing, and the first parachute jump in history was a success.

The successful sky diver, long forgotten, can be identified only as *Canis familiaris parachutensis*.

Kamikaze: Divine Wind

Typhoons, the Pacific counterparts of Atlantic hurricanes, are generally spawned in that vast stretch of open water between Wake Island and the area north of New Guinea. The two types of storm are formed and behave in much the same manner. Also, the typhoon's period of greatest frequency tends to coincide with that of Atlantic hurricanes, with the highest number in June and continuing until the month of December.

Typhoons tend to be the more deadly of the two oceanic storms. Wind-speed measuring devices are generally blown away before maximum winds can be measured, but a full-blown typhoon's winds are suspected to greatly exceed those of killer hurricanes. A hurricane racing in from the Atlantic with winds of 150 miles per hour is always considered extremely dangerous. A similar storm in the Pacific may attain a wind speed of over 250 miles per hour. Why? Typhoons have a much greater area of ocean in which to grow before they are confronted with any large landmass that would reduce their intensity.

Seemingly, the gods of war occasionally employ typhoons as their most effective weapon against an enemy. In 1274 the ever-conquering Kublai Khan, grandson of Genghis Khan, sent his nomadic Mongol troops from northern Korea to invade Japan. A great fleet of about 1,000 ships with 40,000 men headed for the islands. Records of this attempted invasion are not complete, but we do know that the assault never quite got off the ground. A typhoon struck the fleet, sinking about 200 ships and drowning 13,000 troops. The surviving ships were so badly damaged that the invasion was abandoned.

The great Khan tried again in April 1281. This time he struck with one of the largest invasion fleets known in history: nearly 4,500 ships equipped with about 145,000 troops. The attack was launched against Kyushu, the southernmost of Japan's four main islands. The troops, meeting fierce resistance, were forced to reembark on the invasion ships.

The Mongols attacked again in July, and again the battles were so violent that the invaders had to retreat to their ships. They never had a chance for a third try, because a great typhoon descended upon them with most disastrous results. Estimates of the Mongol losses vary, but most accounts agree that at least 4,000 ships and about 100,000 troops were lost. Survivors who made it back to shore were massacred by the Japanese. Thus ended Kublai Khan's ambitions for the conquest of Japan; the Mongols never seriously threatened the island empire again.

The Japanese believed the typhoons were sent from heaven to protect them from their enemies. In honor of the timely interventions on their behalf, the Japanese named the typhoon "kamikaze," meaning "divine wind." The Japanese adopted this name during World War II for their suicide pilots who deliberately crashed obsolete planes, each carrying a 550-pound bomb, into Allied naval vessels. The kamikaze pilots did much damage to the U.S. fleet, at the small price of about 2,000 of their bravest, most dedicated, but apparently expendable

youth. The kamikaze movement evolved out of desperation when it became evident that Japan was going to lose the war.

No doubt, pious citizens of Japan during World War II prayed for the divine wind to strike again at the invading enemy fleet. And occasionally their prayers seemed to be answered, for a devastating typhoon did strike the U.S. Third Fleet during an invasion attempt of the Philippine Islands.

By the mid-20th century, scientists had accumulated an immense body of knowledge about the great oceanic whirling storms. They knew the general location of storm breeding grounds and the conditions under which such activity might be expected. Most of the knowledge was theoretical and therefore of limited use during the war. When it came to outguessing these monster storms, the encounter between the U.S. Third Fleet and a typhoon in mid-December 1944 clearly demonstrated that a little knowledge could be a very dangerous thing.

The Third Fleet, under Admiral William F. Halsey, was to participate in the invasion of the Japanese-held Philippine Islands. On Sunday morning, December 17, 1944, 90 ships of the fleet's Task Force 38 began refueling from navy oil tankers. The sea was so choppy that the ships were ordered to calmer waters to avoid accidents. Obviously a major storm was brewing, but no conditions substantiated the typhoon idea, so meteorologists assumed that the unpleasant weather was by now northeast of the fleet and moving away. This information was relayed to Admiral Halsey, who then ordered the fleet to turn south to escape the storm even more quickly. That tragic change in course, which was based on flawed meteorology, made disaster inevitable. It led the task force directly into Typhoon Cobra (as it was later named). Midmorning on December 18, Halsey's flagship *New Jersey* encountered two dreaded signs in quick succession: the barometer began a rapid fall, and the winds shifted to the north, then to the west of north. The meteorologist aboard the flagship realized that such a counterclockwise change could mean only one thing—a typhoon. He also knew that the information he had given the admiral earlier was wrong, that he had erroneously estimated the storm's location.

As fate would have it, Halsey's Third Fleet blundered into the center of the typhoon. Destroyers, cruisers, battleships, and carriers were tossed about like corks. The order came out of the flagship to "proceed at will!" It was every ship for itself, and the fleet scattered wildly before the wind and waves. Spread out over 2,500 square miles, all except the largest aircraft carriers and battleships were in trouble.

On the small escort carriers, chaos was the order of the day. The ships bucked so wildly that planes tore loose from their restraining gear and crashed about the flight decks, smashing into one another and occasionally catching fire. The crews labored heroically to push the loose planes overboard. In all nearly 150 planes were destroyed by the storm.

The worst agonies, however, were suffered by the fleet's destroyers. The little "tin cans," prized for their speed, were virtually unmaneuverable despite their 40,000-horsepower engines. The surviving skipper of one destroyer recalled later how his ship, the *Hull*, heeled over into the pounding seas. He tried every combination of rudder and engines to right the ship, but it was rolling 70 degrees. The 130-mph wind forced the ship to lie steadily over on her starboard side until the seas came flowing into the pilothouse itself. The ship then remained on its starboard side at an angle of 80 degrees or more as the water flooded into the upper structures and the sea poured down the ship's stacks. All the captain could do was to step off the bridge into the water as the ship rolled over on its way down to a watery grave. Of the 264 men on board, only 62 survived. The *Hull*, a survivor of the Pearl Harbor attack, was one of three destroyers to go down.

A crewman from the sunken destroyer *Monaghan* remembered the horror that occurred below deck when his destroyer heeled over, lay heavily on its starboard side, and gradually settled. As the sea rushed in, he and about 40 other men tried to open a door, which was now overhead, to escape. After a desperate struggle they forced it open against the wind and, one by one, wiggled out. Emerging on deck they inflated their life jackets before being washed into the boiling sea. Some of the men were pounded into a pulp against the side of the ship; others were swept away. The crewman felt as if he were in a whirlpool. Men were knocking against him as he started for the surface, and he could feel them grabbing at him. Someone had managed to inflate a life raft, and several of the men scrambled aboard. Pitching and tossing, as much under water as on it, the little group clung to their raft as the typhoon gradually subsided. Later, in calmer seas, they watched in helpless horror as circling sharks fed on their shipmates' bodies. Altogether 82 sailors from the three destroyed ships were picked out of the heaving seas, but 790 were gone.

The destruction wreaked by Typhoon Cobra seemed to suggest that the divine wind had once again protected the Japanese from their enemies. Not a single ship from the Third Fleet escaped major damage, and three destroyers sank, with the loss of most of the crews. Later a court of inquiry blamed the disaster on Admiral Halsey, since he was the responsible commander. Many scholars believe the devastation of the Third Fleet was the inspiration for Herman Wouk's novel *The Caine Mutiny*.

In an ironic reprise, six months later Halsey was preparing to support the invasion of Okinawa when Typhoon Viper approached the area. Once again the admiral, following his weather expert's advice, took his fleet on a course intended to escape the storm, and not surprisingly he ran right into it. Six men and 76 planes were lost, and 33 ships were damaged. The divine wind seemed to have a special vendetta for Admiral Halsey. But this was the last time the forces of the kamikaze would come to the aid of Japan. Even the supernatural was helpless against the atomic destruction of Hiroshima and Nagasaki, and the war ended.

Monster Born in a Dream

In medieval Jewish folklore, a robotlike servant shaped from mud to resemble a man was magically endowed with life. It was called a *golem*, which originally meant "matter without shape." According to the Talmud, Adam was created in seven stages, beginning with the collection of dust and continuing through a shapeless mass (golem). He was complete, ready to stand on his feet, when he received a soul.

The golem could be kind and virtuous but, because it was without a soul, was unable to do anything except follow the instructions of its master. This it did, in a most literal way, often inadvertently causing great destruction. The golem was activated by a charm called a shem, a paper inscribed with one of the names of God. When the shem was placed in the mouth or inserted in the head of the inert mass, it possessed the ability to move about and obey commands.

The most famous of these imaginary creatures was the Golem of Prague, said to have been created by the Rabbi Judah Lowe in the late 16th century. Lowe used the creature as his weekday servant, but on Friday afternoons he removed the shem that gave it life so that it could rest on the Sabbath. One weekend the rabbi neglected to take out the charm, and the golem ran amok; after Lowe caught it and removed the shem, the creature crumpled to dust. The main purpose of the rabbi's golem had been to protect Jews from anti-Semitic violence. With the help of his golem, the rabbi was very successful in bringing criminals to justice and exposing anti-Semitism. He once discovered just in time that the Passover matzoth had been poisoned.

Another golem, created by a Rabbi Jaffe in Russian Poland, was given the tasks of lighting fires and performing other duties not permissible to Jews on the Sabbath. On one occasion this creature far exceeded its orders and burned up practically everything in sight before it was destroyed.

The golems of Jewish mythology also served as an inspiration for a 19th-century writer's masterpiece of horror. So fascinated was the writer by the golem that one night she dreamed of this soulless monster wandering abroad. Her imagination took over as she recognized the tragedy of a life with no guidance, hope, or purpose.

Outside a gloomy Swiss chateau the night was wild, split by several storms that raked the landscape with forked bolts of lightning. Inside the chateau an 18-year-old girl sat among friends by the flickering fire. She listened with fascination as the men in the group discussed the evolution theories of one Erasmus Darwin (father of Charles Darwin), and she joined their conversation when it turned to their common interest in the supernatural.

The friends agreed that before retiring for the night they would each write a ghost story, but the girl, feeling the need for sleep, promised to write her story the next morning. During the night, she was awakened by a nightmare so vivid that she stayed awake. She was so shaken by the experience that she feared sleep lest the same nightmare return.

At the first light of dawn she began to write down her dream: "By the glimmer of the half-extinguished light, I saw the dull yellow eye of the creature open; it breathed hard, and a convulsive motion agitated its limbs." And so was recorded the birth and awakening of Frankenstein's monster.

The year was 1816, and the girl was Mary Wollstonecraft, who later became the wife of the English poet Percy Bysshe Shelley. The chateau was the Swiss home of exiled poet Lord Byron. Two years later her nightmarish dream was published as *Frankenstein, or the Modern Prometheus* (since Prometheus had also attempted to create a human from mud).

The Gothic novel recounts the story of Frankenstein, a medical student who builds a human body of parts gathered from dissecting rooms, butcher shops, and cemeteries. He animates the creature, but because it lacks a soul the monster of Frankenstein, like the golem, is responsible only to its creator. Moreover, as often happened with golems, it breaks out of his control. Longing for sympathy and shunned by everyone, the creature ultimately turns to evil and brings dreadful retribution and eventual destruction on Frankenstein for creating him.

Hollywood has taken many liberties with the only noteworthy book written by Mary Shelley, and many films—some good, more bad, others both funny and frightful—have re-created the subject matter. The creature, called the Monster by Shelley and identified as "It" by its maker, is often confused with its creator and is popularly called Frankenstein. The story ends with the Monster addressing an explorer after it has killed Frankenstein and all of his loved ones. It explains that Frankenstein, who created a man without love or friend or soul, was actually the real monster.

CHAPTER TWO

The Great Uranium Rush

A great uranium rush occurred in the 1950s, as many people searched for the proverbial pot of gold at the end of the rainbow.

Uranium, a radioactive metallic element used chiefly in nuclear technology, is very toxic both as a chemical and as a radioactive substance. It has the highest atomic number of the elements found naturally on the earth. Elements of higher atomic number are man-made. Many scientists believe the element is the product of the radioactive decay of elements of higher atomic number that existed at an earlier time. Possibly those parent elements were born from supernovas or other such stellar events; perhaps they were created during the "Big Bang" that marked the beginning of the universe. Though they do not exist on earth, there is no reason to believe these parent elements do not exist elsewhere in the universe.

The first and most spectacular use of atomic energy was in the pursuit of modern warfare, to terminate World War II. Since then, major countries of the world have developed incredible nuclear arsenals that, if put to use, would end most life on the planet.

The great potential for nuclear power lies in nonmilitary, peaceful uses. A nuclear plant first generated electricity in 1952, and since then much power for industrialized nations has come from nuclear reactors.

Other applications include propulsion for submarines, ships, and space vehicles; excavating tunnels and canals; and producing power for remote areas such as Antarctica. Until the hazards of nuclear power, principally those related to safety of nuclear reactors and disposal of radioactive wastes, are adequately solved, its development will move slowly.

The United States exploded its first atom bomb in New Mexico on July 16, 1945, thus inaugurating the atomic age. The race for development of nuclear energy was on. The first problem, to the chagrin of U.S. government scientists, was the shortage of uranium ore for research and development. The Colorado Plateau of Utah and Colorado was known to be abundantly supplied with undiscovered hoards of uranium. This was particularly true of the Utah deserts, although major deposits also existed in Colorado, Arizona, and New Mexico. To provide an incentive, the government offered a generous bonus of $10,000 to anyone who discovered and mined radioactive uranium.

The incentive worked. The government launched a uranium rush far more energetic than the 1849 California gold rush. More prospectors scoured the Utah deserts during the mid 1950s than had ever invaded California during the rush for gold.

Two special features made the search for uranium slightly easier. First, uranium minerals were highly radioactive, so any self-respecting prospector would be armed with a Geiger counter or scintillator. Either instrument would emit a series of lights and beeps that would immediately alert the prospector if he was on the right track. The other feature that helped the uranium hunt, particularly in Utah, was the color of the uranium minerals. Although many varieties of ore minerals yield uranium when refined, the chief mineral in the Utah search was carnotite. Its bright canary yellow color could be seen for great distances in the bright sunlight.

Most uranium ore deposits are sparsely distributed. Each typically contains less than 0.3 percent concentration of the mineral. Searchers think they have found their "pot of uranium" when they discover the first small pocket of uranium minerals. Unfortunately, thousands of these uranium mineral pockets are needed to make an ore body that can be mined at a profit. Many prospectors on the Colorado Plateau did find Geiger counter–responsive carnotite and had their excitement dampened by discovering that it was not in economic quantities.

Prospecting for uranium became, for Utah residents, a weekend trek into the desert. (The people who made the most money from the 1950s uranium boom may have been those who sold prospecting equipment and clothing.) The Utah deserts are spectacularly beautiful at any time of the year. On a clear day the view, undistorted by moisture in the atmosphere, seems to go on forever. Distances are deceptive, and what appears to be only a mile or so away may in fact be 5 to 10 times that distance. This is what the Utah deserts are today, but many changes have taken place in Utah since the time, in the Jurassic Period of 150 million years ago, when the great masses of uranium were being deposited.

Utah, 150 million years ago, was a lush, tropical, humid land similar to the Amazon River Basin of today. The area was basically a lowland of dense jungles and glades crossed by an abundance of muddy streams and dotted with ponds, lakes, and swamps. Plant-eating dinosaurs hid fearfully beneath protecting foliage while huge carnivorous dinosaurs crashed and thundered through the forests searching for prey. Uranium prospectors of the 1950s, spending months and years on the relatively barren Utah desert, would find the above description of this land as it existed during prehistoric times next to impossible to believe.

The fortune hunters concentrated their search for the yellow minerals in the many ancient stream channel deposits. Organic material could be found in great abundance, especially if the deposits had been laid down in a tropical environment. A major method of preservation is volume-for-volume replacement of organic matter with inorganic matter. Usually groundwater carries the minerals that will petrify the preserved material. The most common is silica, but occasionally the groundwater also carries valuable mineral matter such as uranium. In the Utah desert, the uranium mineral carnotite was one of the fossilizing materials of the organism. As a result, many of the fossils found on the Colorado Plateau of Utah are somewhat radioactive, including ancient trees and bones of some dinosaurs. One petrified log, over 100 feet long and 4 feet in diameter, yielded more than 100 tons of uranium ore valued at $230,000. It was, beyond doubt, the most valuable log ever discovered.

In July 1955, a geologist discovered an almost complete skeleton of a juvenile stegosaurus. Encased in a layer of volcanic ash, it had been killed in an eruption that also killed all the scavengers that would have fed on its remains. Despite its value as a scientific treasure, the stegosaurus met the same fate as the petrified log. The results, however, were less impressive, for it yielded a minimal harvest of uranium.

Successful prospectors often tried to mine the deposit themselves rather than sell it to professional mining companies. Few novices, however, were aware of the dangers of radioactivity. Because most of the mining was done underground in well-defined shafts, uranium minerals surrounded the workers in the walls, ceilings, and floors of the shaft. They were constantly bombarded with deadly radioactivity. The number of miners who gave their lives in their quest for riches will probably never be known.

One such event involved three brothers who struck a very rich deposit of carnotite near Cortez, Colorado. Since all three were mining engineers, they were able to set up an efficient mining operation that was producing only high-grade ore. So rich was the deposit that at the end of a day's work the brothers emerged covered with yellow dust and were amused at the value of the residue they washed away. Little did they realize that each moment they spent underground took them closer to disaster, for they were constantly subjected to concentrated radioactivity from all directions. The yellow dust they took home simply ensured that the work hazard was still with them. In less than two years the three brothers had become quite rich; they were also quite dead of radiation poisoning, leaving behind three extremely wealthy widows.

By the end of the 1950s the U.S. government was well supplied with uranium ore. With the bonus incentive discontinued and government demand for the product fully satisfied, the great uranium boom came to a screeching halt. All that remained was a supply of treasure tales to add to the folklore of the West. When refined, carnotite yields two metals: uranium and vanadium. One of the best-known stories that circulated on the Colorado Plateau concerned a miner who had prospected around the turn of the century. He was excavating a carnotite mine for its vanadium content. The miner was unable to extract vanadium at a profit because the ore had too much of the almost "worthless" uranium. He abandoned the mine but kept its location a secret. When the boom of the 1950s began, the miner realized his good fortune and set out to reclaim his old mine. Unable to relocate the mine, he became a familiar figure to the prospectors—an ancient battered, weatherworn miner leading an equally weather-beaten mule. His disappointment must have become unbearable, for one morning a party of prospectors found him slumped over with a bullet in his head.

A Harness for the Wind

About 10,000 years ago a group of Late Stone Age people established a campsite near the shore of a now-extinct lake in the vicinity of Yorkshire, England. That these early campers were hunters is indicated by numerous remains of game animals. Scientists believe that they also fished and even traveled on the lake. Excavation of this ancient site uncovered a wooden paddle, the oldest and most primitive implement of water navigation to have been discovered anywhere in the world.

Perhaps a Paleolithic ancestor invented the paddle while he still navigated by straddling a log.

A more efficient means of water travel would require something that could provide more speed and control with less human labor—such as a sail. The Nile River extended the entire length of ancient Egypt. At the delta, the only place where the country widened, the Nile split into at least seven arms interconnected by numerous stream channels. People were drawn to this web of waterways at a very early date. Egyptians, not surprisingly, became key contributors to the history of water transportation. The earliest known record of a sail is a picture on an Egyptian pot dating from about 3200 B.C.—over 5,000 years ago.

King Cobra

A panel of herpetologists (scientists whose specialty is snakes) was asked which snake they considered the most dangerous in the world. They almost unanimously agreed that the king cobra, *Ophiophagus hannah*, ranks number one. This was the snake that experts would least like to step on, to be locked up with in a phone booth, or to see smiling down on them.

The king cobra is not the most deadly. The saw-scaled viper, taipan, and krait all have more toxic venom; the mamba is faster; and the gaboon viper has longer fangs. Moreover, several of the 300 species of small, peaceable sea snakes are 50 to 100 times more venomous than the cobra but are the least threatening of all poisonous snakes. The shy sea snake usually bites only when handled by a foolish human. Recently a scuba diver off Thailand suddenly found himself in the midst of an incredible number of sea snakes. They were so thick he felt that he could walk on them; yet not a single snake made any attempt to bite the man who, at this point, was mentally writing his will.

The king cobra is by far the largest venomous snake in the world, its record being 18 feet, 4 inches long, with an average length of about 14 feet. It is found throughout Southeast Asia, India, and southern China, but it is not a major contributor to snakebite mortality worldwide—which reflects its preference for regions remote from the habitats of humans. If it were to invade agricultural areas as do other cobra species, the bite and death ratios would shoot up significantly. Scientists estimate that the king cobra, with large poison glands containing highly potent neurotoxic venom, can deliver 120 times the amount of venom needed to kill an adult human. Moreover, to make sure it delivers an ample amount, the king cobra hangs on when it bites, chewing away at the wound so that the venom penetrates. The venom is so toxic

that careless handling of the substance can send a person into a coma. Being toxic to the nerves, king cobra venom can kill a person within 15 to 20 minutes after a bite.

The king cobra can also topple an elephant! In 1991 a timber crew in central India was startled when one of their work elephants suddenly trumpeted in agony and went berserk. It attacked everything standing, including other elephants. Much destruction resulted from its 20-minute rampage, after which it suddenly stopped, sank to its knees, and gently rolled over dead. The unfortunate beast had unwittingly stepped on a king cobra that protested much as any creature might do when stepped on by a full-grown elephant. After biting the elephant on its right front knee, the snake died from the trampling and became a tasty evening meal for the workers; so did the elephant.

As its Latin name implies, the king cobra is a snake eater (*Ophio*—snake, *phagus*—eater). Although its diet is almost entirely snakes, it does not hesitate to swallow lizards, rodents, and birds that wander into its habitat. It will eat any snake, harmless or poisonous, including pythons and cobras, and has been known to turn cannibal and eat the young of its own species.

One of the king cobra's claims to superiority, demonstrated by those studied in captivity, is its intelligence. For example, it quickly learns not to strike the glass cage front and can recognize the person who cares for it. It is therefore tolerant toward its keeper but will become aggressive to others who come too close. Visitors to a zoo notice how its eyes glitter with brilliant, round pupils and how its stare is so intense as to be frightening even from behind glass. A strain of stubbornness also marks this species of snake in captivity. In the Los Angeles zoo, a 12-foot king cobra went on a hunger strike, for reasons unknown, and had to be force-fed. A crew of keepers kept it immobilized while the curator shoved food down its throat.

One of the greatest dangers of the king cobra is that it is so unpredictable. It may move quietly away from an intruder or disturbance, but all too often it is aggressive and will attack without provocation. Without doubt a 14-foot poisonous snake is a force to be reckoned with. When excited, angered, or threatened, it raises the front one-third of its body off the ground, extends the anterior ribs so that the hood flares menacingly, and gives a prolonged hiss. This familiar cobra stance is probably the serpent's way of warning a potential enemy. Most intimidating is the king cobra's ability to move forward while in this upright posture, with its head four to six feet off the ground. The terrifying head, at eye level or above, can strike as far as the raised portion of the body can reach. Challenged by the formidable gaze and stance of the cobra, all adversaries should retreat first and evaluate the situation later.

The king cobra is more predictable during breeding season: both sexes are guaranteed to be aggressive. Both guard the eggs and challenge all intruders. The king cobra deserves commendation as a most devoted, sacrificing parent, for very few snakes nest in this manner. After mating, the female, occasionally assisted by the male, will scoop piles of vegetation into large mounds. She will cover the eggs with vegetation and make another compartment on top of them. Into this she will crawl and remain until the eggs are hatched. The male, if he decides to stick around, will station himself on either side of the mound, changing position from time to time. The decaying vegetation keeps the eggs warm, as does the body heat of the attending female. When the eggs hatch, the new hatchlings are on their own, and the parents go their separate ways. The newborn cobras, about 20 inches long when released from the egg, are already a deadly threat to animal life and are very excitable. They strike at just about anything that moves, even each other.

During the incubation period, the guarding mother is most dangerous. Even the male knows better than to climb the mound, as the female will not tolerate any disturbance of her incubating offspring. Any unfortunate human who stumbles onto the cobra's nest without recognizing it will not be given the advantage of the cobra's hiss and stance; he or she will be immediately attacked and bitten. The hapless victim will probably have no trouble sleeping through the next night or, for that matter, eternally.

The Year of the Popes

Petrus Hispanus (Peter of Spain), born in Lisbon, Portugal, in 1215, was a prominent physician during the Middle Ages. He had studied in Paris and later taught at Siena, Italy. He was the author of a number of medical treatises, including the important and popular *Treasury of the Poor* (*Thesaurus Pauperum*). This medical compendium, written so that common folk could easily understand it, offered remedies that were easy to obtain and use. For example, the author recommended lettuce leaves for a toothache, lettuce seed to reduce the sex drive, and the topical application of pig dung to stop nosebleeds.

The writings of Petrus Hispanus were widely read, and his medical treatments and advice were much sought after. Some of his treatments were based on his own discoveries, such as the application of sulfur to cure scabies. Others merely confirmed the medical practices of the 13th century, an age when remedies seemed to be valued according to their unpleasantness. For a woman in a hysterical faint he recommended that the best therapy was to blow salt and pepper up the patient's nose and she would come around promptly.

Although he was the source of many valuable speculations, Petrus Hispanus, as did his medical contemporaries, believed in demons and witchcraft. He advised epileptics to carry a parchment inscribed with the names of the Three Wise Men, and anyone who desired to be popular and wealthy to wear the heart of a vulture. In 1276 Petrus Hispanus was appointed physician to the Vatican, and he ministered to Popes Gregory X, Innocent V, and Adrian V. Under his undivided attention, the three popes died within seven months. The Church ignored the coincidence and, because the prime requisites for the next pontiff were youth and good health, Petrus Hispanus became Pope John XXI. His knowledge of hygiene, his medical skills, and his prediction of a long life for himself were sufficient guarantee that he would last longer than aged and decrepit cardinals.

Unfortunately, neither youth nor good health nor medical skill was an effective defense against a prophecy of longevity that goes awry. Within seven months of his ascent to the papacy, Pope John XXI was killed when the roof of the palace that had been built to his specifications fell on him. His death came as a relief to orthodox clerics, who considered him to be a heretic, an Antichrist hostile to God and the Church, and a man of science at a time when popes were not supposed to be distracted by scientific thought.

By some medieval logic, the death of Pope John XXI and three other popes within 14 months seemed to carry a warning to the Church to avoid electing a physician to the papacy. To this day no other pope has been a man of medicine.

Firestorms

Whirlwinds are, by definition, swirling masses of air that originate at ground level and are never extensive. They generally occur in the summer when warm land surfaces heat the air rapidly, triggering cyclonic activity. Usually local, they rise only a few hundred feet. Observers will comment that "it looks like a tiny tornado."

When the whirlwind is spawned by fire, it becomes an entirely new kettle of stew. Tornado-like whirlwinds are sometimes formed when extreme heat is generated quickly, as in volcanoes and fires. The heat sends scalding air rushing upward, and the typical tornadic whirling action begins. One volcanologist watching the 1943 eruption of the Mexican volcano Paricutín recorded seeing a great rush of gas from the crater. It formed a miniature tornado about 20 feet in diameter that roared away as if propelled by some irresistible force.

On September 1, 1923, the Tokyo–Yokohama area of Japan was devastated by an earthquake that brought extraordinary destruction by flooding and fire. The final horror came from the "dragon twists" (Japan's descriptive name for the fire-induced whirlwinds). The intense heat generated from burning Tokyo sent hot air rushing upward at an estimated speed of over 150 miles per hour. It formed an immense pile of artificial cumulus clouds that actually extended higher into the atmosphere than Mount Everest. Survivors of the great earthquake watched horror-struck as writhing pillars of black smoke, skyscraper-high, whirled down upon them in true tornado style. Bodies were tossed in all directions as blazing human torches. The spectacle resembled a monumental fireworks display. In one sector alone, a single tornado fire whirlwind was estimated to have killed over 40,000 people!

Artificially induced whirlwinds can occur in forest fires, such as occurred on August 23, 1951, at Vincent Creek, Oregon. Firefighters noticed that, in the midst of violent whirling surface winds, a dark tornado-like tube was extending upward over the fire. The top was obscured by drift smoke at approximately 1,000 feet. The tornado winds were so extreme that a green Douglas fir about 40 inches in diameter at breast height was twisted and snapped off about 20 feet above the ground. Near this fire-whirling wind the flames leaped several times higher than those in surrounding areas, and treetops burst into flames like the flash of a powder keg as the fire whirled by. The fire tornado whirlwind disappeared and reappeared rapidly at least three times during a 10-minute interval.

People in bombed-out cities during World War II were no strangers to such whirlwinds. Fire raids were terrible enough, but when these raids turned into dreaded firestorms, all human effort to escape them was useless. During an air raid over Hamburg, Germany, on July 27, 1943, the rain of incendiaries and high explosives, and the fires started by them, turned the sultriness that had gripped the city into violent heat. The result was a firestorm that burned out eight square miles of the city.

Fires everywhere were suddenly linked together, heating the air above to such an extent that a violent updraft occurred. This caused the surrounding fresh air to be sucked in from all sides to the center of the fire areas. The terrific suction caused air movements much more powerful than normal, so that they became fire tornadoes. And, as is typical of tornadoes, the resultant funnels twisted large trees out of the ground, tossing them as giant torches into the air. Clothes were torn off people fleeing for shelter, and cars and trucks were overturned. To complete the devastation, the oxygen at ground level was burned up, and thousands suffocated in the bomb shelters. Truly, war was never more comparable to hell.

One of the most mind-boggling and disastrous firestorm catastrophes occurred in the United States, yet it is one of the least remembered and recorded of natural disasters. The incredible event was eclipsed by another catastrophic event that happened at the same time.

In 1871 a pair of gigantic firestorms drove flaming whirlwinds through a number of towns in Michigan and Wisconsin, killing more than 1,000 people. Autumn of that year was extremely dry in the Midwest, with no significant rain after July 8. Near the border of Wisconsin and Michigan small blazes plagued the loggers and sawmill operators because the dense pine forests along Green Bay were constantly victimized by careless hikers and campers. The local people, ever alert, managed to subdue most of the blazes.

Almost inevitably, in the face of so extreme a drought, on October 8, 1871, an ominous yellow veil of smoke hid the sun over Peshtigo, Wisconsin. The residents of this small lumber town had no way of knowing that a sudden rising wind had caused several scattered blazes to unite into a single conflagration that was advancing on them like an army of solid fire.

Shortly after 9 P.M. a fearful roaring came from the southwest, and a massive firestorm struck the town. Houses crumpled like paper; flaming roofs were borne away like gigantic fireworks. The air itself became so incendiary that people simply died by inhaling it, and hair and clothing exploded into flames. The people crowded into the Peshtigo River by the hundreds, fighting for space to submerge themselves between breaths. Several of the town merchants lowered their merchandise into wells, hoping to save it, and then lowered children on top of the goods. To their unspeakable chagrin, the merchants had actually condemned the children, because within minutes the goods would ignite, burning them to death immediately.

One man who recorded some of the happenings called this catastrophic wind a "tornado of fire." He described it further: "When I heard the roar of the approaching tornado I ran out of my house and saw a great black balloon-shaped object whirling through the air over the tops of the distant trees." What the man had seen was a whirlwind spawned not by weather but by fire. Scientists now know that large, intense blazes can create massive updrafts and surface winds with hurricane force. In turn, these winds spawn powerful vortices and become tornado-like whirlwinds!

Over 750 people died that day in Peshtigo. But it was not the only town to suffer so. On the other side of Green Bay another fiery tornado swept through the Door Peninsula. Between the two fires 23 towns and villages were devastated and hundreds of isolated farmsteads leveled. Authorities placed a conservative estimate of deaths at over 1,500, and at least 1.3 million acres of timberland were destroyed.

The event came to be called the Peshtigo Horror, yet news of this catastrophe did not rivet the attention of the nation, and it is still practically unknown. Why? On that very same evening, several hundred miles away, Mrs. O'Leary's legendary cow knocked over a lamp in its barn and set the building ablaze. Although the Peshtigo Horror resulted in over five times as many deaths, it was always well hidden in the shadow of the Great Chicago Fire.

Perhaps some day the film industry will make a movie of the Peshtigo Horror as they have done numerous times about the Chicago Fire. It would make a truly gripping disaster story.

The Ultimate Conversation Piece

About 70 million years ago, a pregnant *Hypselosaurus* positioned herself to deposit her eggs in the nest she had built to house her future family. As she laid her brood one at a time, her 40-foot body shuddered each time an egg dropped into the nest. How many eggs she laid or whether any of them actually hatched into infant hypselosaurs will never be known. What *is* known is that one of the eggs did not hatch. It remained intact, with an encased embryo that probably died shortly after the egg was laid.

The egg was able to escape scavengers and agents of decomposition, so it had probably been covered with mud and buried, immediately and permanently. Groundwater that carried chemicals in solution gradually replaced all of the organic material of the egg with inorganic substances, mostly silica, so the egg became petrified. In 1930, when the intact egg was unearthed by a farmer as he plowed his vineyard in Aix-en-Provence, it created quite a stir.

This area of southern France was famous for its yield of dinosaur remains. Many shell fragments had been found in this locale as early as 1869. The discovery of a fossil egg in such good condition helped to confirm that the earlier fragments, along with the egg itself, were indeed from a hypselosaur. The only apparent damage that time had inflicted on the egg was that the shell contained many cracks. Otherwise it was a solid chunk of rock that had once been a dinosaur egg of the genus *Hypselosaurus*.

Such an old fossil in excellent condition is only half the story. In 1992 the prestigious Christie's of London auctioned the fossil egg unearthed in southern France. Never had there been such an unusual sale. The egg went to an anonymous buyer for the incredible bid of $11,000. Asked why anyone would pay so much for an unhandsome, useless trinket, the spokesman for the new owner explained that the egg was purchased for the sheer novelty of owning it: people who see it would react with astonishment at the very idea of a 70-million-year-old egg.

Somehow, an egg that old, one that could contain five and a half pints of liquid, would seem a bargain at any price.

Isn't Love Grand?

Amor omnia vincit! The ways of love among animals are as different as the ultramicroscopic virus is from the 180-ton blue whale. To humans many animal rituals appear bizarre, corrupt, deviate, ridiculous, kinky, and sometimes counterproductive. But each species is guided by a sure instinct provided by nature's impeccable logic. Regardless of the manner in which it is accomplished, the primary function of all animals on earth is to perpetuate each species.

Perhaps no member of the animal kingdom is so self-sacrificing as the male praying mantis. He is no doubt aware of his mate's reputation for insatiable cannibalism; yet the mating instinct coerces him to go ahead and take his chances. One mantid, for whom biting off the head is an essential part of the mating ritual, is most cautious. At his first glimpse of a female, the male will freeze. Then, creeping up to her in extremely slow motion (one-eighth of an inch in one hour), he carefully keeps to the left, out of her line of vision, so that she won't devour him prematurely.

Finally, if he has judged the distance correctly, he springs at her and lands in the proper position for mating to begin. Unfortunately for him, a neural mechanism within his head inhibits the release of sperm, so the head must be destroyed in order that the eggs be fertilized. This happens automatically when the female bites off his head. Because nerve centers of the mantis are located in various parts of its body, the headless male can walk, raise his wings, and proceed with the mating game. The female continues to feed on him, absorbing essential proteins, until mating is complete. Often only his wings remain.

In most species of frogs the male does not chase after the female; he merely gets her attention, and the female comes to him. When the mating season approaches, the male frog sits in a stream or pond with other males, usually in a large assemblage. They all blow up their cheeks and croak, the result being a cacophony of grunts, bellows, squawks, trumpets, hoots, honks, caterwauls, and yodels that cannot be ignored. Apparently the noise captures the attention of female frogs, for they will head for its point of origin.

Many male frogs croak in a chorus in a well-organized, highly refined din. In several species of frogs, scientists have observed a "choral master" who leads the croaking. One can almost imagine the warty conductor standing on a rock podium that overlooks the choir of males, waving a stick and instructing, "All together, boys!"

Once the females arrive, mass mating hysteria begins. So carried away is the male that he will, without hesitation, grab and leap on whatever happens in his path—a stick or stone or even another male—and try to mate with it. Or several males may jump onto the same female, or a male leaps to piggyback on another male who is busily riding a female. The otherwise-engaged male seems unaware of the burden on his back.

In the midst of all this mating frenzy a male may grab a lump of mud, which being soft and malleable feels much like a female. The mud is modeled in his tight grasp until it assumes the modified shape of a frog. The male, thus deceived by his creation, may hold on for days waiting in vain for the captive lump of clay to release spawn. Despite the mud-loving male and other hit-or-miss frog matings, in the 300 million years that frogs have been on earth many species have obviously developed successful, albeit unique, methods of spawning.

Most of the 2,800 species of dance flies (family Empididae) are murderous predators. The male, being smaller than the female, instinctively knows he must protect himself from his cannibalistic mate. He may distract her by offering an edible marriage gift or a "toy" of sufficient interest to keep her occupied while he mates with her. The males of several dance fly species bring the gift to the female wrapped in a beautiful package. With spinning glands at the tips of their forelegs, they spin a fine white silken thread around a captured insect. The female accepts the packaged gift and busies herself with unwrapping it.

The males of some species are overly cautious in their deception. The tricky male will capture a very tiny insect (and what he considers tiny is almost microscopic from the human perspective). Around this minute morsel the courting male will spin a gigantic silken balloon, often twice his size. The female readily accepts this enormous package and straightaway sets about the arduous, time-consuming task of unraveling it. By the time she reaches the infinitesimal gift within the bigger-than-life package, the male has consummated his marriage and moved on. The ultimate in deceptive packaging is achieved by the male who weaves a silken blimp and presents the oversized gift to a favored female who unravels and unravels, only to discover long after the male has mated and fled that there is nothing inside!

The males of the species *Hilara sartor* are probably responsible for the common name "dance fly." They do not try to deceive the female with worthless, ephemeral, or nonexistent gifts. Rather each spins a beautiful white veil that he spreads out between his four posterior legs. The veil in no way suggests food, but when courting males gather together and appear to dance in the air with the sun shining on them, they lure hundreds of females. Viewers describe the scene as if tiny elves and mystical fairies are dancing in the golden sunlight, twisting and turning in the gentle breeze. The glowing white veils attract females just as the gift parcel attracts their less ethereal cousins. The female *H. sartor* accepts the bridal veil from the male and lands in the grass clutching her gift. While she plays with the veil, the male mates with her and then leaves abruptly.

The loligo squid, which includes several varieties from 4 to 10 inches long, would escape our attention entirely but for their spectacular mating experience. At mating time, a calm sea will suddenly appear to be boiling as the squid gather spontaneously in unbelievable throngs and participate in sexual orgies that may last up to four days. The mass of frenzied creatures with a single purpose grows constantly in numbers as more and more loligo arrive on the scene.

The living tide of squid becomes incredibly dense, and all of them are obsessively undistractable. They are particularly vulnerable at mating time, and the sea creatures and birds for which they are an important food are quick to take advantage of the windfall. Sharks can move through a mass of squid with their jaws wide open, gulping and swallowing the abundant harvest nonstop. They feast without diverting a single squid from the orgy at hand. Scientists observing a rendezvous of squid have estimated several million of them in a volume of water 400 feet in diameter and 10 feet deep, about 80 ten-inch squid in each cubic foot.

The first and only mission of each squid in the great gathering is its frantic search for a partner. In the manic mob confusion, most individuals attach themselves to the first available squid, male or female, hurriedly separating unless an opposite-sex squid joins them. Many are embracing in pairs, their tentacles intertwined; others are dancing face to face, and numerous masses of writhing tentacles include five or six individuals. By the second night more actual matings and fewer nuptial dances are occurring. Color changes accompany the increased erotic frenzy as the squids dance and intertwine. Normally delicate in color, iridescent and almost translucent, the male squids become an intense purple, and their heads and tentacles are striped in red and maroon during the embrace.

Actual mating takes place when the male loligo squid transfers spermatophores (torpedo-shaped tubes filled with sperm) from his mantle cavity to the glandular patch under the female's mantle. His left fourth arm is modified to scoop and deposit the tubes. This he repeats several times before relaxing his embrace. The pair then sink gradually to the bottom, where the female deposits gelatinous capsules of eggs, anchoring them to the sea bed. In three days the females deposit 10 to 20 capsules, arranging them in bunches that resemble chrysanthemums. The male, making doubly sure of fertilization, emits semen into the water surrounding the eggs.

The number of eggs produced by a squid orgy is colossal. The egg capsules may cover 200 acres of seabed. By the time the females have deposited all of the eggs, they are exhausted to the point of death; many of them will be eaten by the somewhat fatigued males. In the end most adults from the rendezvous of squids will be corpses among two billion to seven billion eggs.

The egg capsules appear unprotected, but the covering hardens and its very unpleasant taste renders them acceptable food for none but the least discriminating predators. Within a month the eggs will hatch, and new squids, small but recognizable, will emerge into the sea to become an immediate food supply for predators. Those that escape begin a life that will climax with its predictably violent, frenzied, agonizing finale.

Mating and fertilization in bedbugs is involved, difficult, and bizarre. To begin with, the female has no external opening in which to deposit sperm. The male triumphs over this minor inconvenience by making his own entrance; on the fourth abdominal segment of the female is a notch that marks the spot. Here, far from the female's reproductive system, the male punctures a hole into a mass of tissue that protects her internal organs from being lacerated. He discharges a large quantity of semen into the female, and the sperm make their way through the blood to sperm reservoirs.

After the female takes a meal, the sperm migrate to the ovaries to fertilize eggs as they are formed. On fertilization the eggs are extruded from the opening made by the male. The opening heals over, but scars from each penetration remain. Therefore the entomologist, and possibly the male bedbug if he's interested, can determine how many times the female has mated simply by counting her scars.

Mating and egg-laying for the bedbug can occur whenever the temperature is high enough. On their preferred host, humans, the temperature is almost always controlled to their prescription. The rate at which the eggs hatch also depends on the temperature and how recently the female had fed before she laid the eggs; it can take from one to six weeks. The larvae cannot molt on their way to becoming adults until after they have found food. Tiny, white, and transparent, the bedbug larva can be seen clearly only when its body is engorged with blood.

Gift-giving, particularly of food, has a variety of meanings and purposes among animals. The male may attempt to seduce the female with the gift, or he may give her something to eat so she won't eat him. He may be offering a bribe that will demonstrate his superiority over others, or he may be showing simple, practical evidence of his ability to provide for her during her confinement.

The northern terns are species of birds that live in a nonaggressive, cooperative community. Though the male tern does not have to seduce, distract, bribe, or impress the female, he does reassure her with the ritual presentation of food. The offering of a superior-quality fish is his way of testing whether the female likes him enough to mate with him. His gift helps her to assess his potential as a provider during the egg-laying and incubation period.

The male tern ready to mate will catch an especially handsome fish and carry it back to the breeding colony. There he struts back and forth

among a group of eligible females, holding his head high and thrusting out his chest. As he parades he slyly inspects females for one worthy of his attention and his gift. Finally he will offer his fish to one of them. The female's assorted reactions may include ignoring the male completely, turning her back on him, or looking critically at the fish and walking away bored and annoyed. She may even take the fish in her beak for a moment but must return it promptly if she doesn't intend to mate with the owner. To keep the fish for even a brief time means that she is pledged to be his mate, despite any lack of charm or delicacy of the suitor or his gift.

A male tern who has been turned down by numerous females may hawk his wares to the entire colony. Then, if no female responds to the offer he will, almost in defiance, eat the fish himself and fly off to catch a more attractive fish to offer to another group of eligible females. He will continue the ritual until he has successfully found the female tern of his dreams.

The female that does accept the fish shows no sign of eagerness to eat it. She simply holds one end of the fish in her beak while the male, possibly as a sign of emotional harmony between them, holds the other end of the fish in his beak. The male and female just stand quietly alongside each other, staring straight ahead, each of them loosely holding on to opposite ends of the fish. This test of mutual generosity may last an undetermined length of time.

And so, when the moon comes over the mountain, its silvery light may shine on two very weary birds, a male and a female, standing alongside each other, staring straight ahead and still holding on to opposite ends of a smelly old fish!

Seismic Cows

Many animals are able to feel vibrations at much lower energy levels than humans can. Scientists now know that some animals can feel earthquake vibrations long before humans even become aware that the ground is shaking. This provides a method of earthquake prediction, and much research on animal reactions to seismic activity is being carried out at leading institutions of higher learning.

Animal reaction to an earthquake preceding human awareness has been known for a long time. In March 1933, a young couple driving home to their farm located just outside of Long Beach, California, were amazed to see their cows all lying flat on the ground, with heads outstretched in a wheel-like pattern. The couple stepped into the circle and tried to nudge the animals to their feet, but to no avail. The cows rigidly remained in this protective circle. A few minutes later the disastrous 1933 Long Beach earthquake struck with all its fury.

More recently, on August 6, 1979, the San Francisco peninsula was rattled by a moderate quake that added more credibility to animal prediction of earthquakes. Minutes before the quake struck, caretakers noted that zebras, ostriches, deer, and antelope, which normally intermingle in the African veld, separated themselves by species, each grouped and huddled together. It was as if the species were angry with each other. Within one half hour after the tremor, all animal behavior returned to normal as if nothing had happened.

Land of the Roc

Eons ago the island of Madagascar was part of Africa. Ancient geologic upheavals caused it to split from the continent and, in the course of about 80 million years, to "drift" to its present location approximately 250 miles east of Africa. Being a large, isolated island Madagascar developed unique flora and fauna. Among its specimens is an earthworm a yard long and an inch thick. Here is also the home of the tenrec, a hedgehoglike mammal acclaimed in the Guinness Book of Records for the largest mouth of any mammal its size and for the largest number of young born in a single litter (32, although the normal litter is a modest 12 to 18). The island is the home to several small primates, such as the aye-aye and the lemur. The largest carnivore is the catlike fossa, two and one-half feet long.

Small though the typical fauna of Madagascar may be, it was also the home of the legendary roc. Herodotus, in the fifth century B.C., wrote of a race of huge birds described by Egyptian priests. Found beyond the source of the Nile, these birds could carry off men in their talons. During the Middle Ages a boatload of Arab sailors that had landed on Madagascar told tales of a gigantic bird. Beyond its size, their descriptions were quite limited, so imagination was used to fill in the blanks. The stories became part of the *The Thousand and One Nights*. The tale of the roc (or rukh to the Persians) was narrated by the long-suffering Sinbad the Sailor. According to his account, he arrived at "a beautiful island abounding with trees bearing ripe fruits, with birds warbling, and pure rivers." Here Sinbad spied a huge, puzzling white dome. It appeared to be a large building so, in search of an opening, he walked completely around it—a full 50 paces.

While contemplating the smooth, impenetrable dome, Sinbad noticed that the sky had darkened. Looking up he beheld a gigantic bird, most surely the roc, whose young were fed elephants. The parent roc would prepare an elephant for eating by seizing one in its claws, flying with it to great heights, and dropping it. Then the roc would carry off the smashed carcass to its nestlings. "And while I wondered at the works of God . . . that bird alighted upon the dome, and brooded over it with its wings, stretching out its legs behind upon the ground; and it slept over it." Sinbad realized then that the enormous dome without an entrance was the giant egg of the giant bird.

Madagascar was also visited by Marco Polo, who may have actually seen the great bird. His description was as unrestrained as Sinbad's. He apparently told Kublai Khan about the roc, so the Khan sent an expedition to Madagascar in 1294 to confirm the story of the bird "so big in fact that its wings covered an extent of 30 paces, and its quills were 12 paces long and thick in proportion, and it feeds on elephants." The expedition brought back only a feather from the roc. But what a feather! It measured "90 spans in length" (about 60 feet).

Scholars believe that the Khan's men probably brought back a frond of the raffia palm that, when dried, superficially resembles a feather. The Khan doubtless believed, much to the relief of the sailors, that this was a feather from the fabled roc. He rewarded the leaders of the expedition quite handsomely.

Considering the actual observations of the Arab sailors and of the notoriously reliable Marco Polo, both evidently saw a bird of tremendous size. Evidence also shows that an enormous bird lived on Madagascar through the 14th century and became the inspiration for the legend of the roc. Further documentation shows that this bird existed until about 250 years ago but is now extinct.

When France took possession of Madagascar in 1642, authorities noted that some of the natives kept water in enormous egg shells. A history of the island by the French governor referred to a giant bird in the south that laid gigantic eggs. In 1850 a French merchant obtained three eggs and some roc bones. He shipped them safely to Paris, where they were studied by the awestruck director of the Paris Zoo, Geoffrey Saint-Hilaire. He reported on two eggs measuring 13 by 8½ inches and 15⅝ by 12⅝ inches, the largest bird eggs ever seen; each held over two gallons of fluid. One egg was capable of holding as much as the contents of six ostrich eggs or 150 chicken eggs. An omelet made from a single roc egg would serve 75 people.

Saint-Hilaire imagined a bird resembling an ostrich, powerful and flightless, that stood 16 feet tall. He named it *Aepyornis maximus*, "the tallest of high birds." In 1866, in a swamp, a scientist named Alfred Gandidier discovered some huge bones in a perfect state of preservation. At first sight they seemed to belong to some large pachyderm, but they proved to belong to *Aepyornis*. It was at once nicknamed and is still known as the elephant bird. Near the end of 1866 a complete skeleton of *Aepyornis* finally came to light in Madagascar. Instead of being as tall as a two-story building it was a mere nine or ten feet tall. It must have weighed almost a thousand pounds in life; being four times as heavy as an ostrich, it was hardly an inconspicuous bird.

The mystery of the roc was finally settled to the general satisfaction of those who wondered about this monstrous bird of the vast eggs that had been glimpsed in the jungles of Madagascar and had inspired legends. Unable to fly, *A. maximus* could scarcely have carried off an elephant or blotted out the sun as it spread its wings. These are pardonable exaggerations of storytellers; the real bird was impressive enough.

All the indications are that the elephant bird died out about 250 years ago; there is no evidence that the island's human population caused its extinction. They seem not to have hunted *A. maximus*, although they may have helped its demise along by gathering the eggs.

Scientists have inferred that *A. maximus* was a resident of jungle swamps. Hundreds of years ago the climate of Madagascar began to change, becoming gradually less moist. This caused many of the swamps to dry up, and the bird would thus have been driven into an ever narrower habitat, finally unable to gather enough food or find the right shelter to survive. Many *A. maximus* skeletons have been discovered in a large dried mud deposit of a former swamp. Researchers believe this was the area in which the last surviving members of the species huddled together until death overtook them. There were no survivors.

In far-off New Zealand was a bird that would have towered above the elephant bird of Madagascar. It was the moa, which actually did stand 12 feet tall. But that's another story.

Don't Try This at Home

Diamonds *will* burn. This important, or at least interesting, information should restrain impetuous, inquisitive people who might otherwise consider burning their diamonds to find out for sure.

That diamonds will burn if properly ignited has long been known, but no scientists knew just why. Most people simply could not imagine that a mineral so hard and that looked so much like glass (which everyone realized would melt but not burn) could be consumed by fire. So in 1797 an English chemist named Smithson Tennant allowed his curiosity to get the best of him. He just had to know what diamonds were made of, so he borrowed his wife's largest diamonds and put them to the test by burning them. Tennant carefully collected the gas given off by the burning diamonds and chemically analyzed these fumes. He found that they were carbon dioxide.

Tennant's analysis proved that despite the diamond's glassy appearance it is nothing more than a crystalline form of almost pure carbon—the same common material of which coal, graphite, charcoal, and other combustible materials are made. Tennant had made a great discovery, but his wife was more than slightly provoked that he proved the chemical constituents of diamond at her expense. So she left him.

Stone Age Baseball

Nolan Ryan, who retired from baseball after the 1993 season, pitched in the major leagues for over 25 years. He holds the major league record for strikeouts (over 5,000) as well as for pitching no-hitters (seven). At age 45, he still threw a fastball over 100 miles per hour!

Should someone ask when the fast pitch was invented, the logical answer would be, "A few days after baseball." However, increasing evidence indicates that our early human ancestors may have perfected a fast ball—or fast stone—that even Nolan Ryan might envy.

The French explorer Count de la Perouse recorded that, during his travels through the South Pacific in 1787, he sent a water-replenishing party ashore on the Navigators Islands, now known as Samoa. He lost 12 of the 61 men almost immediately when they were caught by surprise in a barrage of rocks thrown so hard "they produced almost the same effect as our bullets, and had the advantage of succeeding one another with greater rapidity."

This is certainly not the only record of primitive peoples employing the fast pitch with baseball-sized stones to besiege an enemy. British historian J. G. Wood wrote in *The Natural History of Man* (1870) that Australian aborigines occasionally killed gun-toting British soldiers with stones. Dancing crazily from side to side so that the soldier would be unable to take deliberate, accurate aim, the aborigine would unleash "a shower of stones with a force and precision that must be seen to be believed." We know now that many of the Pacific island natives employed the fast stone as a weapon of war and hunting, and many of them could have put the greatest of modern baseball pitchers out to pasture with their speed and precision.

With Wood's book as a guide, scientists searched museum collections for these early baseball-like weapons. A number of handstones were found gathering dust in drawers simply because the curators, though they recognized the stones as artifacts, had no clue to their use. Most of the handstones were lemon shaped, which suggests that they were thrown with a spin, rather like miniature footballs.

In the well-known biblical story, David defeated Goliath with a missile hurled from a slingshot. History fails to emphasize that many soldiers in early skirmishes were equipped not with bows and arrows or spears but with fist-sized stones carried in quivers or pouches. They were extremely efficient and skillful, and a barrage of hurled stone missiles would mow down an attacking force more quickly and effectively than an assault with arrows would. The soldiers were also equipped with short swords for close-in fighting with any enemies that survived an attack of stones.

In the late 1980s the remains of a soldier were uncovered in the vicinity of the walled city of Jericho. He was a casualty of the stone-throwing infantry. His shoulder had been broken by the impact of a hurled stone, and his skull fractured by either a club or another thrown stone. He had died clutching a stone the size of a baseball but oval in shape. Clearly he was preparing to hurl his missile when struck down. At his side was a pile of seven throwing stones that had lain in his now decomposed missile pouch. The implication is that warfare always ended with hand-to-hand combat. The scientists who uncovered this warrior's skeleton nicknamed him the "lone soldier of fortune."

Examination of fossil bones found at early-human sites of two million years ago, such as Olduvai Gorge in Tanzania, reveal scratch marks made as the hominid feasted on the meat. Since all of the stone implements discovered here were for cutting, not fighting, the question remains—how did these very early men hunt animals? Undoubtedly some were obtained by scavenging, but the problem still remained of how to keep other scavengers away from the kill.

The answers lay for many years at the feet of scientists excavating the sites but were ignored. These sites are abundantly littered with smooth, roundish stones about the size of baseballs. They would not have been suitable for flaking into tools but were ideal for throwing. Rounded stones found in great quantities at early kill sites were the hunting weapons of the time. One such stone was found embedded in an early hominid's skull, demonstrating that they were also very effective implements during times of conflict. The very nature of the fractured skull indicates the tremendous force with which this missile was thrown. It was definitely a killing blow, for the physical evidence indicates that this was not a postmortem assault. The fact that the stone was never removed suggests that the skull itself may have become a war trophy.

The hands of early humans were suited for the hurling of stones. As far back as 3.4 million years ago, *Australopithecus afarensis* (Lucy's people) could shape their hands into the three-fingered grip that baseball pitchers now use for strength and control of the ball. Apparently stone throwing comes so naturally that some anthropologists believe this could have been one of the incentives for our ancestors to walk erect. One thing is certain: the fastball pitch was used for hunting and defense at least 3 million years before the mid-19th century, when baseball was officially supposed to have been invented!

The lethal reputation of the hurled ball was reaffirmed in August 1920. Shortstop Ray Chapman of the Cleveland Indians became the first and only on-field fatality when he was hit by a pitched ball just above his left ear. He died one day after he was struck, and shortly thereafter the batting helmet was invented.

Sometime in the future a pitcher of incredible skill, strength, and stamina may exceed the record of Nolan Ryan. We must hope, however, that no hurler of the hardball will be able to surpass the killer record of our primitive ancestors.

Pig Soldiers

During the third century B.C. the Romans issued picturesque copper bars to commemorate pigs. The Romans were not inclined to commemorate casually; when they did, it was for very good reason. In this case they considered the honored pigs to be thoroughly deserving.

The army of Caesar was under heavy attack by the army of Pyrrhus (319–272 B.C.), whose use of war elephants was a very effective offense. The Romans, completely unprepared for this type of warfare, were being pushed into an imminent defeat, and the footsoldiers fought with real desperation.

The din of battle was overwhelming to a large group of pigs being kept in the Roman encampment as a ready supply of food. The noise so unsettled them that they panicked and tore down the fences that enclosed them. As the pigs ran helter-skelter between the feet of the advancing war elephants, the huge beasts also panicked. Uncontrollable, they reared up, throwing off the riding soldiers along with the mahots (drivers) and fled the battlefield in complete disarray. With them went Pyrrhus's army.

The Romans achieved complete victory but in a manner so unexpected that the army of Caesar just stood and watched dumbfounded as the enemy fled in total confusion. For years afterward, to eat a pig or, for that matter, to harm one was forbidden. In recognition of the "service" of the pigs, the Roman government issued commemorative copper bars to honor the unofficial pig soldiers.

Shark Myths Dispelled

At the beginning of World War II the U.S. Navy issued survival manuals to naval and army personnel traveling on troopships. The manuals contained information guaranteed to make the readers shark wise, or at least to dispel fear of sharks. Unfortunately, readers were told that

sharks are slow-moving, cowardly, and easily frightened by splashing, although it is almost common knowledge that sharks are very fast swimmers, they fear and retreat from nothing, and splashing only attracts their attention. The manual further advised that someone caught in the water with a shark should first knife the shark, then immediately "swim out of the line of his (the shark's) charge, grab a pectoral fin as he goes by, and ride with him as long as you can hold your breath." With advice that appears to have been assembled by a hungry shark, it's almost a miracle that the United States won the war.

That many troops confronted by sharks took advantage of this manual of shark advice is doubtful. However, two airmen downed in the Pacific did use the manual. They had inflated their life vests and were in the process of tying themselves together so they would not drift apart. To multiply their uneasiness, they saw a large shark fin circling them about 30 feet away. In desperation one of the men threw the shark manual at the predator as it closed in. The shark devoured the manual with apparent relish. As it chewed, it seemed to assimilate the book's contents, for it swam away without bothering the airmen (the very shark behavior explained in the book!). Less than an hour later the airmen were picked up by a rescue ship.

In 1958 a group of 34 scientists from several countries met in New Orleans to consider possible ideas for developing an effective shark repellent. One result of this meeting was the development of the Shark Research Panel, an assortment of scientists committed to researching shark behavior and especially to studying shark attacks. The U.S. Navy, in cooperation with the Smithsonian Institution, worked closely with this group of scientists. Over a period of nine years they logged and analyzed some 1,600 attacks, the earliest of which involved a seaman who had fallen overboard in 1580. Records suggest that his demise somewhere between Portugal and India was exceedingly gory. The study dispelled some long-held myths about sharks.

One very popular belief is that the presence of porpoises in the water means that no sharks are about, since the porpoise is one of the few creatures sharks seem to fear. In reality sharks and porpoises frequently inhabit the same water but, for the most part, tend to respect and ignore each other. Another myth, thoroughly exploited by television, movies, and books, is that a porpoise will defend a human from an aggravated shark attack. This is purely a pipe dream. When an attack is in progress, the porpoise wisely swims off to avoid the possibility of being in the way of a frenzied feeding session.

Writers and illustrators have tantalized readers with images of triangular fins circling a hapless swimmer as sharks prepare to charge and mutilate in a mass attack. Most attacks, however, are made by a single shark that the victim rarely sees before being struck. Although the victim may be bleeding badly, other sharks in the vicinity usually do not participate. The prey is the property of the original attacking shark. Investigators are beginning to believe that human blood is not the strong stimulant it has been presumed to be, but a string of bleeding fish on a diver will immediately attract a shark's attention.

Massive shark attacks do occasionally occur, usually after a major maritime disaster. As one might expect, throngs of sharks are quick to intrude in waters around naval battles such as those that took place during World War II. Several surviving sailors on rafts recall with helpless horror the ocean's surface dotted with circling shark fins that engaged in a frenzy of feeding on dead and dying bodies floating on the water. Unfortunately, this is not a myth.

The next myth, a very popular one, can be disregarded. Most people imagine that fatal or maiming shark attacks involve a monstrously huge shark in deep water far offshore. A statistical investigation of shark attacks shows that about two-thirds of them take place in water little more than waist deep, and in about half of these incidents the shark is less than six feet long.

On February 26, 1966, in the middle of the afternoon on the south coast of New South Wales, a 13-year-old boy was wading about 30 yards from shore in less than five feet of water when his right leg was seized by a shark. He tried to beat the creature off and in desperation even leaned over and bit the shark on the snout. Still it held on, and the boy's screams resulted in the ringing of the shark warning bell.

Lifeguards arrived on the scene but could see no shark—just a very frightened boy standing in shallow water. One waded out and tried to pull the boy ashore but couldn't budge him. The man felt along the boy's immobile leg until he touched the shark. In record time he withdrew his hand and signaled his associates for help. It took six men to drag the boy ashore with the shark doggedly attached to his leg. The jaws of the shark, an eight-foot great white, were pried loose from the boy on the beach. The boy survived the loss of excessive amounts of blood but walked with a slight limp for years. As for the shark that bit him, its days were over.

Some definite conclusions emerged from the systematic study of shark attacks. One is that attacks occur, not surprisingly, in direct proportion to the incidence of sharks and humans sharing the same waters. Most of the near-shore attacks are carried out by the great white and tiger sharks that frequent the shoreline. They are the true villains, but most of these sharks are relatively young and therefore not very large.

The white-tipped shark is suspected of being the principal predator of people forced into the open sea as a result of some type of maritime disaster. The circling shark fins observed by surviving sailors after deadly naval battles were probably those of white-tipped sharks.

In the 1950s abrupt and unpredicted attacks on scuba divers made headlines. The great white sharks in the chilly, murky waters off California were the offenders. On July 14, 1959, just off San Diego, two men were diving for shellfish in about 24 feet of relatively clear water. Suddenly one of the men began to yell for help and went under. The companion looked down and through his face mask saw his friend wreathed in blood, protruding trunk first from the mouth of a 20-foot white shark. This was the last time the attacked diver was ever seen.

Attacks have continued to the present, although these same sharks had almost ignored humans before that time. The change in behavior does not seem so far-fetched when one considers that scuba diving became a popular sport in the 1950s. Great white sharks, who have been feeding off the California coast for thousands or even millions of years, can easily mistake people in black wet suits for seals in these dim cloudy waters. Scientists believe that most shoreline attacks are really a mistake in identity; but once it latches on to a human, the chance of the shark's just spitting the victim out is very unlikely.

In 1964 a commercial fishing boat dredged some fossil shark teeth off the coast of Maine. The teeth were over four and a half inches wide at the base and just slightly over four inches long. Teeth of this huge sea monster, *Carcharodon megalodon*, have been dredged up many places in the world, particularly off California. *C. megalodon*'s length, based on the teeth, could have reached over 60 feet and its weight more than 50 tons; it was certainly big enough to swallow a small truck. This species of shark, extinct for about 20 million years, probably satisfied hunger pangs by feeding on whales.

Shark lovers may be comforted by the fact that a near relative is among the living. It is *Carcharodon carcharias*, the great white shark described earlier as the culprit in the San Diego attack. This modern species rarely exceeds 20 feet in length, and its sharp teeth are usually less than two inches long. Recent dredging just off San Francisco Bay brought up shark teeth about five inches long. Shades of *C. megalodon*, except that these shark teeth were not fossil, but definitely recent.

However menacing the image of the killer shark, the odds against being killed by one are 300 million to 1. Sharks, in fact, have far more reason to fear humans than humans have to fear them, because fishermen are now slaughtering sharks at the rate of 100 million a year. This fish, a favorite prey in sportsfishing, is hunted intensely for the grill in U.S. restaurants and drowns in drift nets set for tuna and squid. And every year, over seven million pounds of shark fins are sold to merchants in Hong Kong. According to legend, shark-fin soup imparts sexual, spiritual, and financial powers—to anyone who can afford $50 dollars a bowl.

Shark populations have suffered disastrously from this last myth. In 1991 alone, the fins of over 3.2 million sharks were collected for shark-fin soup. After its dorsal fin is harvested, the shark is usually tossed back into the sea; shark fins are 30 times as valuable as shark fillets. Unable to swim, the unfortunate shark soon starves to death. Because shark populations are being decimated worldwide, authorities are trying to curtail fin harvests. And their argument is sound: starving on the bottom, tiger sharks or others who eat humans will doubtless crawl to shallow water where they can find easy human prey.

Paleolithic Kitchen Rejects

During the early 1980s, fox hunters from the Indigirka River in eastern Siberia were puzzled that one of their fellow hunters managed to trap a fox every day while they often caught nothing at all. They investigated and found that the successful hunter had been baiting traps with meat from a fossil mammoth! The foxes apparently were fascinated by the unusual taste of well-aged meat that had been preserved in the permafrost for thousands of years.

Scientists who heard the news explored further and eventually found traces of an ancient human settlement. It was dated radiometrically and found to be at least 13,000 years old. The investigators found many charred mammoth bones that they originally described as "kitchen rejects."

Later, when the scientists found meat still covering many of the bones, they reasoned that the bones had been buried by the ancient hunters and saved for a rainy day. The rainy day apparently never arrived, and the buried cache was eventually forgotten. The permanently frozen condition of the encasing soil served as a refrigerant, and the meat had been perfectly preserved during 13 millennia.

The fossil cache of kitchen reserves was accidentally discovered by the modern fox hunter, who decided to try some as bait. The success far exceeded his most optimistic hopes. Although tempted, he never screwed his courage to the point where he would taste the ancient meat himself.

Today's Oldest Living Things

A little over 7,000 years ago bacterial spores were embedded in mud lining Elk Lake in present-day Minnesota. Scientists from the United States Geological Survey collected samples of this mud in 1988. In the laboratory they warmed the spores to surface air temperature and put them in a nutrient-rich culture where, to quote one of the scientists, "They grew like crazy!" After 70 centuries they were still alive; they had survived by living in a frozen state of suspended animation. The scientists were sure they had discovered the oldest living things. And they were right, at least for a short time.

In 1992 scientists discovered what could be the largest and nearly the oldest living thing on earth. The organism is a giant fungus, an interwoven filigree of mushrooms and rootlike tentacles spawned by a single fertilized spore nearly 10,000 years ago at the close of the last ice age. It's come a long way from that single spore, for it now extends over more than 30 acres in the soil of a forest near Crystal Falls, Michigan, along the Wisconsin border.

The fungus *Armillaria bulbosa* is genetically uniform from one end of its expanse to the other, which is why scientists say it rightly deserves to be called a single individual. If all its mushrooms and tendrils are considered together, the fungus weighs about 100 tons, about as much as the more compact blue whale. The organism survives by feeding on dead wood and other detritus, spreading outward right beneath the surface as it senses the presence of nutrients nearby.

To quote one leading fungus researcher's comments on this discovery, "The catchy part of it is, when you really begin to appreciate how large this thing is, it's mind-boggling. People usually think of a mushroom as a little creature, but most of the action of a fungus is underground." It is almost the world's oldest living thing but not quite.

The Michigan mushroom probably got its start several years after a California creosote bush sprouted in the East Mohave. This complex growing in Soggy Dry Lake has colonized the area with a circle of bushes 25 meters (over 81 feet) in diameter. Any creosote bush that has become established in an arid area captures moisture so efficiently that no other plant can grow near, not even another creosote bush. The individual bush takes over the area by sending out new stems around its base, to be nourished by the spreading underground roots. All are part of the single organism. The current circle of creosotes suggests that this single organism has been growing and expanding for over 10,000 years.

Centuries ago an ailing mastodon stumbled onto the grounds of what is at present the Burning Tree Golf Course, located about 25 miles east of Columbus, Ohio. The mastodon came there to die. Perhaps the moisture of what is still a peat bog felt soothing to its aching body as it lay down and soon breathed its last.

The carcass of the mastodon was slowly engulfed in the bog and buried by subsequent deposits of silt and clay. Much of the body was preserved in its original state. Eons later, in 1989 to be precise, workers excavating the peat bog to create a small lake between the 11th and 15th tees uncovered the remains of the fossil mastodon.

Authorities, hurriedly consulted, excavated the mastodon's remains scientifically. In the abdominal region of the skeleton were the remains of the mastodon's last meal. It consisted mainly of swamp grass, leaves, small branches of pine trees, and seeds. The plant material was traced back to the time of the last ice age. Radiometric dating of the wood from the mastodon's meal indicated an age of about 11,500 years.

Enclosed in the woody material near the ribs was a reddish-brown cylinder of malodorous material. Back in the laboratory, analysis of this foul-smelling, decaying material revealed it to be tens of thousands of bacteria. These single-celled organisms were identified as two strains of *Enterobacter cloacae*, bacteria commonly found in the intestinal tracts of mammals to aid in digestion. They were cultured by the scientists because, amazingly, they were still alive!

After searching for, and not finding, the bacteria in 12 soil samples from the excavation site, the scientists determined that they did not come from the surrounding soil. A second, "blind" analysis by another laboratory in Columbus confirmed these negative results. The bacterial strain definitely came from within the mastodon. The bacteria had survived because they were sealed under clay and sediment. The bog water entrapping them slowed down the bacterial growth enough to cause them to go into a state of suspended animation.

At the moment that these minute bacteria awakened from their 11,500-year nap, they became the oldest living things. Until another organism, most likely a bacterium or fungus, is discovered in some airtight bog or layer of permafrost that is more ancient, these *E. cloacae* will continue to hold the geriatric record.

CHAPTER THREE

Flu, 1918–1919

"The year men cried..."

A Milwaukee nurse

The epidemic struck as the fighting of World War I neared its end, and flu displaced war as the sorrowful centerpiece of everyday life. About two billion people were living on earth in 1918, and as many as half of them fell ill with the flu.

In just over a year, from March 1918 through April 1919, it spread worldwide in several waves and killed an estimated 22 million people—almost three times the number killed in the "war to end all wars." Including the people who died of complications, the toll was probably as high as 30 million deaths. To quote an article in *Science* in 1919, "Never before had there been a catastrophe at once so sudden, so devastating, and so universal."

Over 500,000 Americans died of this pandemic, which began as a mild incident on the morning of March 11, 1918, when a mess cook with chills and fever reported for sick call at Fort Riley, Kansas. He complained that his head and muscles ached and his throat felt sore.

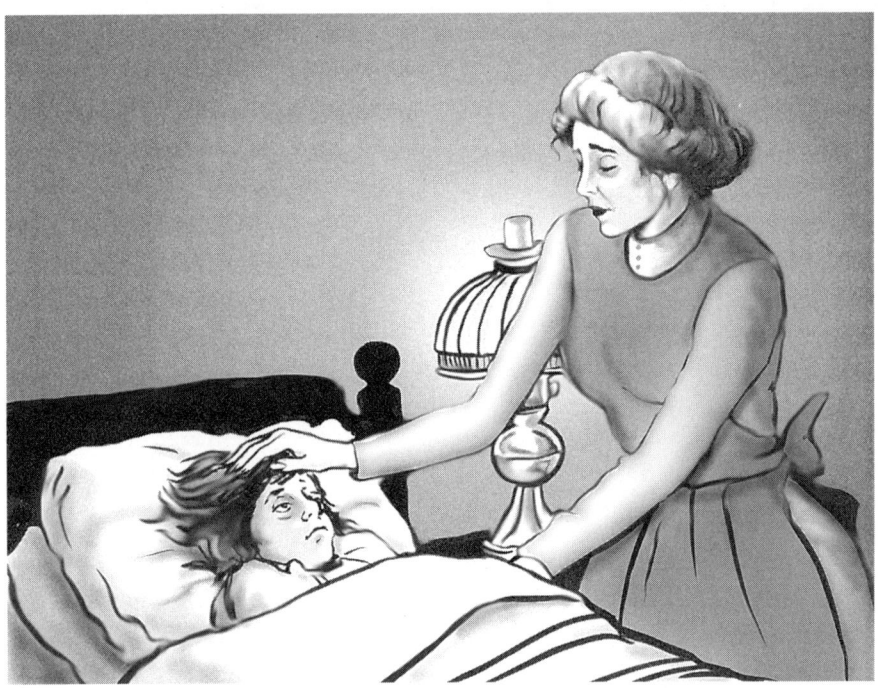

By noon 107 men had been admitted to the hospital with influenza. The caseload grew to 522 by the end of the week, and within five weeks there were 1,127 cases and 46 deaths. Doctors were puzzled and disturbed because the flu was hitting not the very young, very old, or infirm but robust people in their 20s and 30s, those whose resistance to disease is normally high.

Historical pathology lists at least 94 epidemics of influenza between A.D. 1173 and 1875. Fifteen were pandemic, affecting a high percentage of the world population. Hippocrates first noted a probable epidemic in Athens in 412 B.C. Centuries later the Italians christened it *influenza di freddo*, which means simply "influence of the cold." Later the belief grew that some sudden shift in the atmosphere could mysteriously communicate the disease to entire areas.

Fort Riley had experienced a notable atmospheric change just before the outbreak. Gale-force winds had shrouded the camp with a noxious pall of prairie dust, aggravated by stifling smoke from burning horse manure. The United States was up to its elbows in plans for winning the war, so the military chiefs were willing to accept this simple, finite explanation.

The deaths at Fort Riley and other crowded training camps were attributed to complications from pneumonia. So the army continued training two million men and shipped them off to France. Often they were packed into airless freighter hulls that were floating test tubes for breeding virulent, improved mutations of a flu virus not yet identified. Doctors searching desperately for the offending microorganism would discover many years later that the only constant in the flu virus was change.

U.S. troops landing in France by the summer of 1918 brought plenty of flu with them. Within a short time, flu began to provide the war with some stiff competition in the killing fields and occasionally proved superior. General von Judendorff blamed the flu for halting Germany's victory drive in July.

The pandemic spread quickly to England and to Spain, where it killed eight million Spaniards. It became known, in recognition of its swift and deadly invasion there, as the Spanish flu and was nicknamed "Spanish Lady," even though it had originated in the United States. Within four months, flu encircled the world. By August 1918 the disease had changed to a far more vicious form. Once again it spread around the world, and this time vast numbers of people died primarily from subsequent pneumonia, for which there were neither drugs nor treatment.

In September, when the flu was at its most deadly, General John Pershing, commander of the U.S. expeditionary force, was calling for every available soldier for a massive assault. When the October draft call was canceled because of raging flu outbreaks in army camps and ports, Pershing was almost forced to call off the assault. Although the Meuse-Argonne offensive did founder for a while as 69,000 troops fell ill, Pershing won the battle—but at a heavy cost from disease and enemy action.

The navy was hit even harder; 5,000 were lost to the epidemic in a service only one-tenth the size of the army. The captain of the USS *Pittsburgh*, a cruiser on patrol in Rio de Janeiro's harbor, saw more than half of his men disabled, and he buried 58 of them. He turned for home, taking the *Pittsburgh* out of the war just as if the ship had been sunk by a torpedo.

The speed at which the Spanish Lady spread was uncanny. It was called the *Blitz Katarrh* (lightning flu) when it reached the ragged ranks of the kaiser's army. The disease then swept into Russia, China, and Japan, down to South Africa, and across to India, where 12 million people died. It then spread into South America.

Horror stories were common. One man in Rio de Janeiro asked another where the streetcar stopped, thanked him politely, and fell over dead. On a Cape Town tram, seven riders, including the conductor and driver, collapsed and died within a three-mile stretch. In the goldfields of Johannesburg, an engine operator was hoisting a steel cage full of miners to the surface from deep underground. As he reached for the brake lever, he died of the flu, and the cage kept going and banged against the lift's overhead frame. Torn loose by the impact, the cage fell back into the shaft, plunging 24 miners to immediate death. In the yearlong pandemic, no place in the world was safe. The natives of South Pacific islands, with no defensive immunity to the disease, were the most vulnerable of all; 20 percent of the population of Samoa died from influenza. Even in as unlikely a place as Nome, Alaska, 176 (59 percent) of the 300 Inuit there became victims of the deadly disease.

The mighty fell along with the meek. Among victims who survived were Kaiser Wilhelm, King George V, King Alfonso XIII, German chancellor Prince Max of Baden, Assistant Secretary of the Navy Franklin D. Roosevelt, British prime minister David Lloyd George, Queen Alexandria of Denmark, and even Mary Pickford in Beverly Hills.

The early heavy toll among U.S. men in uniform was simply the disease's opportunistic response to their being confined in cramped quarters. Not until September 3 was a civilian case documented, in Boston. A week later three men dropped dead on the sidewalks of nearby Quincy. The first death in New York City was on September 15. Then, flu cases, fed by mass movements of uniformed personnel, began popping up all over the country—first by the hundreds and then by the thousands. The toll escalated to over 10,000 domestic deaths in September, and up to 90 recruits a day were dying in Camp Devens, Massachusetts, alone. In Philadelphia, one of the worst-hit cities, 13,000 died from the epidemic by the end of October.

In New York City, the death rate caused a shortage of coffins. The only solution was to carry the dead to common graves in a temporary coffin, from which they were removed and wrapped in a sheet for burial. The coffin was then reused to carry another body. In the last week of October 1918, 21,000 people died in the United States, the highest seven-day mortality toll in the country's history. And in the last four months of 1918, well over 300,000 died.

Americans flailed desperately against an unseen foe. Large gatherings were banned, and almost all public meeting places, including churches and saloons, were closed. In major cities, anyone caught sneezing, coughing, or spitting in public without a handkerchief paid a huge fine.

A worldwide preventive measure was the wearing of face masks. People fixed layers of gauze across their mouths and noses in the hope of preventing the transmission of infectious droplets. Since the virus passed easily through the gauze, the face masks did little good except to discourage contact. Even so, mask slackers were given a rough time, including occasional public beatings. At a West Coast naval base, guards were ordered to shoot to kill anyone attempting to enter or leave without official permission. Drinking fountains were sanitized hourly by blowtorch; phone booths were padlocked or the telephones were drenched in alcohol.

Since there were no medications to treat the flu, anxious people dreamed up their own preventions and cures. Among the prescribed treatments: drink Scotch, eat garlic, remove tonsils, inhale chloroform, sprinkle sulfur in shoes, wear vinegar packs on the stomach, tie slices of cucumber to the ankles, carry a potato in each pocket. In New Orleans, people bought voodoo charms and chanted, "Sour, sour, vinegar V, keep the sickness off of me." One mother covered her four-year-old daughter from head to toe in raw sliced onions. Several men slept with shotguns under the bed, believing the fine steel would draw out any fever present. One physician summed it all up: "It was polypharmacy run riot."

Real heroes of the relentless epidemic were those in the medical profession. Doctors pumped railroad handcars to reach isolated cases and made their rounds by horse, foot, and auto and even in the newfangled airplane. They not only treated entire families but also collected and handed out food and blankets, milked cows, cooked, built fires, heated water, and bathed and dressed the helpless. Doctors ate meals from the family stewpot when they stopped to eat, and fell asleep on the nearest couch. Sometimes they dozed upright in their buggies, warmed by heated bricks wrapped in newspapers as their horses were given free rein. They indeed stretched themselves to the breaking point, and many became victims.

For each person who died of the flu in the United States, an estimated 50 had come down with the disease. All that could really be done was to keep the patient quiet, warm, fed, and medicated against other life-threatening complications.

Fitfully, the long day's night began to fade. People who had survived the disease seemed to be immune from further episodes. Like a hurricane, the mysterious Spanish flu virus eventually blew itself out and seemed to vanish utterly from the planet. No outbreak of the flu before or since has equaled, in size or severity, the great scourge that gripped the world in 1918–19. Although more recent outbreaks—the Asian flu in 1957, the Hong Kong flu in 1968, the London flu in 1972, and swine flu in 1976—have caught our attention, they have been minor pandemics.

Scientists now know that all of this amazing misery stemmed from a virus so small that 2.5 million in a straight line would take up just about an inch in length. Research has established that influenza exists in three major strains, and vaccines can be created to combat all of them. The problem with flu is that each strain is quite unstable and mutates easily, creating a slight variation from earlier forms. This complicates the prevention process, for any new epidemic is almost sure to have originated with a virus changed enough to escape the antibodies still present in the human body from the previous year's vaccine.

The flu continues to live on, but annual vaccine injections keep the deadly type at bay. A stable, reliable vaccine is still an important work in progress. Each year delicate, elderly, and otherwise susceptible people are encouraged to prepare for the flu before the flu season arrives. For arrive it will.

When One Sex Prevails

Scientists have discovered that, during the incubation stage of alligator eggs, a temperature difference in the nest is what determines whether the hatchling will be a male or a female. If the eggs are kept at about 80 degrees Fahrenheit the offspring will be female; if the eggs are kept at 86 degrees Fahrenheit the hatchlings will be male. Since the alligator female is able to control the temperature of the nest in various ways, she is really determining the gender of the offspring.

No scientist believes the mother knowingly prefers male or female offspring. How warm she keeps the nest depends upon many variables: climatic conditions during incubation, the amount and kind of material used to warm her nest, how heavily she piles it on, and so forth.

Many scientists believe this discovery about gender selection in alligators may help us understand the disappearance of the dinosaurs. During the close of the age of dinosaurs, the Mesozoic, the earth was very active tectonically. Mountains were being uplifted, and new ranges, such as the Ancestral Rockies, were coming into existence. This had a profound influence on the world climates. Those climatic changes could have caused one sex of dinosaurs to dominate over the other. This would certainly disrupt reproductive patterns and result in mass extinction.

The Tainted Blue

The phrase "A diamond is forever" is not far from actual fact. Of all substances that occur naturally, a diamond is the hardest, although it is atomically identical with coal and graphite, two of the softest minerals. It is pure crystalline carbon, forged deep in the earth by the most tremendous heat and pressure.

The geometric, very closely packed arrangement of the atoms and molecules that make up the diamond's internal crystal structure is what makes this mineral so hard. Indeed it is 90 times harder than corundum, the mineral of emeralds and the second-hardest naturally occurring substance on earth. The diamond will resist acids of all types and will burn only in the hottest of fires (above 1,400 degrees Celsius). But despite tales to the contrary, it can be smashed with a hammer.

Diamonds are valuable in industry because of their hardness, but their brilliance makes them valuable as gems. A diamond's brilliance is the result of its high refractive powers: much of the light reaching a diamond is reflected back into the stone rather than through it. Although the best diamonds are transparent and colorless, the presence of an impurity may give the crystal a most desirable color. Such is the phenomenon of the unique, awe-inspiring—and infamous—blue Hope Diamond.

According to legend this extraordinary diamond once served as an eye in the statue of the Hindu goddess Rama-Sita but was stolen by a Brahman priest. The irate goddess decreed that bad luck would befall anyone who should wear her eye as jewelry. The fate of the dishonest priest is unknown, but in 1642 pioneer gem trader Jean Baptiste Tavernier brought the diamond to France. He had returned from India with enough jewels to win a barony from grateful King Louis XIV, to whom he sold the 112.5-carat, round, nonfaceted blue diamond. Louis XIV admired it greatly, naming it "The Blue Diamond of the Crown," although it became better known as the "French Blue." He had it recut into a teardrop shape of 67.5 carats.

How Tavernier acquired the blue gem is not known, but reports suggest that the legendary curse followed the smuggler/jewel merchant. He was, according to some accounts, devoured by wolves on the steppes of Russia. Tavernier did die during a trip to Russia, and he may even have been chased by wolves. But since wolves are no longer certified as human-hungry beasts, the legend has a flimsy foundation. More likely, since Tavernier was 84 years old during his last winter in Russia, he may have died of a bad cold.

King Louis, although awed by the French Blue, feared its relentless legend of ill fortune. He is said to have worn it only once, whereupon he contracted a fatal case of smallpox. His successor, Louis XV, completely refrained from touching it and seemed to have escaped its curse. But Louis XVI didn't. Both he and his queen, Marie Antoinette, used the diamond frequently, wearing it at almost all royal functions. The fate of Louis XVI and his queen is well known: they both lost their heads at the hands of revolutionaries.

After the revolution the French Blue, along with other crown jewels, was placed in a loosely guarded glass case. In a famous robbery in 1792, the robbers—if indeed the robbers and the guards were different people—had little trouble carrying it off. The diamond was not seen again for nearly 40 years.

In all probability it was sold in Spain, where it was recut to avoid detection. The Goya portrait of Queen Marie Louise shows her wearing a deep blue diamond cut very much like the one offered for sale in London in 1830. A rounded oval of 44.5 carats, it was later identified as the missing blue. Henry Thomas Hope, a rich British banker and gem collector, purchased it for the sum of $90,000. The gem has been known ever since as the Hope Diamond.

The Hope family added to the bad-luck stories. When Henry Hope died without marrying, a condition considered most unfortunate in those days, his contemporaries blamed it on the curse of the diamond. The blue was willed to his nephew, for whom Henry was said to have very little regard. The nephew, whose wife ran away with another man, eventually became destitute and was forced to sell the diamond in an unsuccessful attempt to stave off bankruptcy. He died cursing his uncle who had forced the deadly blue on him.

By the time the diamond passed out of the Hope family in 1906, the stories that it brought bad luck had become widely known. The gemstone itself almost seemed to be trying to keep up its malevolent reputation. Its next owner was a Parisian gem dealer, Jacques Celot, who committed suicide in 1907. His estate then sold the jewel to Prince Kanitovski, a Russian playboy who was conducting an affaire with a beautiful French actress. He gave her the fateful diamond to wear during one of her performances. In a fit of jealous rage over several men admiring the diamond, he shot her dead in the middle of her act. Two days later he was assassinated, and his killer or killers were never apprehended. It was said the murder was committed by other admirers of the dead actress.

Tragedies continued to haunt successive owners with an almost monotonous regularity until, in 1911, the diamond was purchased for $154,000 by Evelyn Walsh McLean, whose father had struck it rich in the Colorado gold mines. She scoffed at the diamond's curse, had it mounted in a necklace surrounded by white diamonds, and wore it almost constantly. But she lived to see a son killed in an automobile accident, a daughter die of an overdose of sleeping pills, and her husband confined to a mental institution. She died in 1947, lonely and slightly deranged, at age 61. The Hope was purchased from her estate by the New York gem dealer Henry Winston for $180,000. He freely displayed it in a number of shows with no ill effects and finally, in November 1958, donated it to the Smithsonian Institution under somewhat comical circumstances. Winston mailed it parcel post! The postage was only $2.44, but for $145 he insured the package for $1 million dollars.

The Hope Diamond remains at the Smithsonian, a favorite sight for visitors. In a case of bullet-proof glass the blue Hope rests quietly in colorful splendor where all can marvel at it but none can touch it. Not even the guards are permitted contact. And seemingly, since the blue Hope is completely protected from mere mortals who might wish to adorn themselves with Sita's eye, her curse has become powerless. Or is this only temporary?

Kosher Oil

And Zadok the priest took a horn of oil out of the tabernacle, and anointed Solomon.

1 Kings 1:39

The Bible is full of references to the installation of kings of ancient Israel by anointing them with holy oil.

In 1952 a copper scroll was discovered in the Holy Land. In the opinion of scholars, it contained a list of items taken from the temple of Jerusalem by pious Jews and hidden in a number of caves as the Roman army approached the city. As the Jews had anticipated, the Romans did destroy the temple of Jerusalem in A.D. 70.

One item listed on the copper scroll was excavated from a cave at Khirbat Qumran near the Dead Sea. It was a small earthen vessel filled with an oil wrapped in a nest of palm fibers. The excavating team first thought it was a flask filled with dirt, but outside the cave in bright daylight they noticed that a dark oily substance began to seep out. Most components of the oil have been identified by various chemical tests, but one elusive plant source remains a mystery.

It is believed that the Jews hid this bit of holy oil from the invaders because they knew the Romans valued it as a perfume!

It Blends with the Snow

The polar bear is the largest member of the bear family and one of the largest of all land carnivores. The males average seven to eight feet long and weigh 900 pounds, although many are larger; females average 700 pounds. The largest polar bear on record weighed an incredible 2,210 pounds, a white colossus whose wanderings were centered around the North Pole.

The polar bear is a wanderer by nature. Once a land mammal, it is now, in a sense, an animal that followed its mobile food supply to sea and never returned permanently to land. It ranges throughout the frozen Arctic and may, in its lifetime, travel farther than any other four-footed animal in the world. Some males are believed to ride ice floes all of their lives without ever setting foot on land.

Polar bears are strong swimmers, cruising leisurely at six miles per hour using only the front legs and with hind legs trailing. They have been seen swimming over 200 miles from the nearest land, and they appear to know where they are going. Although the nomadic polar bear enjoys a larger acreage than almost any other animal, its range is still confined by the whereabouts of the ringed seal, its favorite prey and about 85 percent of its food supply.

Distinctively different from its brown, black, and grizzly cousins, the polar bear is built for conservation of heat and ease of travel on ice or in water. With a massive but elongated body, a long neck tapering to a narrow head, low shoulders, and a high rump, it appears awkward but moves over the ice with sure-footed grace and agility. Its pelt consists of a thermal blanket of fine hair and a cover of tough guard hairs. Each hair is a hollow tube that traps air, the best of insulating materials. The hairs are actually transparent but appear white because they reflect all visible wavelengths of light. Underneath the pelt is protectively warm, heat-absorbing black skin. A bulky layer of blubber beneath the skin keeps the bear buoyant and warm in near-freezing water.

For most of their adult lives, polar bears are solitary animals. Nomadic males will turn aside from their course rather than confront each other. The mature he-bear is a tyrant that young bears avoid and with whom females consort for only a few days of mating. They will later shun the male, fearing for the safety of their cubs, which they are quick to defend from any aggressor. In general, polar bear society is one of "armed neutrality," in which conflict is restricted to defending cubs, hauling a meal of seals out of water, and fighting for mating privileges. At that time, competition for a female can become deadly.

The only rival to the polar bear's domination of the Arctic is the walrus, two tons of awesome bulk with a pair of formidable tusks. When an encounter does occur, the bear is hardly a match for the great tusks of the walrus and usually comes out second best. Knowing this, the polar bear will wisely avoid such confrontation. However, if very hungry, it will resort to trickery. In 1984 a scientist observed as a bear sneaked up on a full-grown sleeping walrus and brained it with an enormous chunk of ice. The bear fed well on an abundance of walrus that day.

In its search for food, the polar bear uses equally energy-saving devices. Aided by a most acute sense of smell, it lurks downwind from a ringed seal basking in the sun. It will creep delicately toward the seal, its fur blending with the snow and ice. Bears have been observed to push a chunk of ice in front of themselves to avoid being spotted. The seal can see only a chunk of ice and may even wonder why it appears to move. Researchers have witnessed how a bear will approach a seal with a white paw covering its black nose. In a final rush the bear reaches the seal and, with a single blow of its paw, prepares dinner. A seal in the water is an even easier mark, for the bear need only locate a breathing hole and wait for the seal to come up for air. With one movement, the bear plunges its forepaw into the airhole, smashes the seal's skull, and drags it onto the ice.

At times the polar bear may look upon a nice juicy whale as an endless feast. One reportedly attacked a 40-ton whale. The bear leaped onto the whale's back and submerged with it while trying to bite off a chunk of blubber. When the whale resurfaced, the bear was still in the same position, determined to obtain a mouthful of food. On the third submergence, the bear gave up the whale and settled for the chunk of blubber in its mouth. Although the bear is a loner, groups do gather to share the flesh of beached whales. Several witnesses reported observing nearly 100 bears eating one huge whale. Bears entered the carcass

through a hole eaten in the mammoth belly and emerged from the gaping mouth.

The polar bear is virtually defenseless when swimming and can be pushed or "herded" by a small boat in any desired direction. Knowing this, Inuit often drive a swimming bear close to their encampment before killing it. Inuit have always killed bears in order to live rather than to flaunt a useless trophy. The flesh provides food, fat is both food and a source of oil, bones are shaped into various tools and implements, and the hide becomes clothing. Polar bear liver, however, is poisonous because of its high concentration of vitamin A—40 times as much as is found in beef liver. This is a lethal dose, as some early Inuit may have discovered.

Until recently polar bears rarely encountered humans and had no fear of them. Most bears seem now to understand that humans are more dangerous to them than vice versa, and so they tend to avoid people. Contrary to the chronicles of early whalers and explorers, polar bears do not attack humans with hateful fury, but an excessively hungry bear is always a threat. And a wounded bear can be a deadly adversary. An account is on record of an old male bear charging a hunter for over 60 yards, despite the fact that it had already been shot through the heart several times!

In Churchill, Manitoba, polar bears have taken to raiding garbage cans. Although they recycle a lot of garbage, some negative consequences take place. In 1991 an old man went outside to dispose of the garbage and never came back. The son who went to look for him found that the father's tracks had been intercepted by those of a polar bear that evidently preferred fresh meat to garbage. The bear was hunted down and shot. What was left of the victim was recovered and given a proper burial.

Other, less threatening encounters have been equally unnerving. During an expedition to Greenland in 1980, a sailor who went for a walk unarmed encountered a large male bear. It charged immediately. As the terrified man ran, he tore off his jacket so he could run faster. The bear stopped, sniffed the garment, and resumed the chase faster than before. The man continued to remove articles of clothing, and each momentarily slowed down the bear. The man's frantic shouts attracted his shipmates. As they hurried to his aid, the exhausted sailor gave up and remained rooted to a spot. The bear, catching up to him, simply sniffed his hands. He must have smelled good to this bear! When the man's companions showed up, the bear decided this conquest wasn't worth the effort and hurried off. The man, clad lightly in undergarments, passed out.

In 1969 a coastguard vessel in the Canadian Arctic received a visit from an adult polar bear traveling on a drifting ice floe. The crew threw all kinds of food at the bear, who relished the free lunch. But when the food ran out, the bear tried to come aboard much to the crew's alarm. They turned a fire hose on the bear, an ineffective disciplinary action because the bear thoroughly enjoyed the drenching. It raised its paws in the air, apparently luxuriating in the jet of water under its armpits. The coastguard crew finally fired a distress rocket near the interloper, and the bear reluctantly moved away.

The polar bear can be ingenious in captivity. In January 1976 a female polar bear decided to make a bid for freedom at the Lincoln Park Zoo in Chicago. The subzero temperature had frozen the waterfall in her enclosure, so she managed, with much effort, to haul her 600-pound body up the resulting ice bridge. Her freedom was short-lived. Just 100 feet beyond her prison she was met by a keeper with a tranquilizer gun!

Ancient Fossil Fuel

When Marco Polo left Venice he was instructed to bring back samples of every new product he could find. One of the new products he brought back to Europe was coal. He had observed Chinese royalty using coal to heat a room. Astonished by this solid rock that burned, he inquired about its origin. Nobody could answer except that the rock had been burned by their ancestors for centuries.

Through the centuries the Chinese have received the credit for discovering coal and for having used it long before it appeared in Europe. Many textbooks back this up as a statement of fact. After all, didn't Marco Polo bring back the first samples from China? Yes, he did, but recent evidence shows coal was used as a fuel in Europe centuries before it made its appearance in China!

Paleolithic hunters inhabiting eastern Europe lived under Arctic conditions. They constructed stout tents with heating and eating hearths both inside and outside the tents. Scientists excavating these living sites examined the contents of the hearth ashes and found that the fuel used by these ancient people was coal.

How coal was actually discovered by these Stone Age hunters will doubtless never really be known. Possibly log fires were built where there were outcrops of coal, and the primitive people observed and realized that these black rocks would burn slowly and steadily and could not be blown out. Whatever fortunate circumstances brought about some Stone Ager's discovery of coal, it happened nearly 30,000 years ago.

The Sky for a Ceiling

One often hears of many strange events during tornadoes. They sometimes seem to have an almost human knack for playing practical jokes. So it was on a balmy morning in southern Oklahoma in spring 1980. A woman had just stepped into the yard to hang her washing up to dry. Strangely, she did not notice the ominous dark cloud drifting over the area and obliterating the sunlight as the wind began to blow. She had been a witness to this type of weather before, and it meant a tornado was in the making.

As she worked, a large twister descended from the mother cloud and swept up all the standing structures in its path. The woman fled terrified into her house, aware that she did not have enough time to get to the storm cellar a hundred yards away. Instead she took refuge in a closet located under her back stairway. Bracing herself, she could hear the roar of the twister as it passed overhead. The house shook and vibrated; convinced that she was about to die she prayed silently to her Maker. But the roar of the tornado passed, and the house ceased to vibrate. Knowing she somehow was spared, she cautiously opened the closet door. The woman recoiled at the sight in front of her; all that remained of the house was the closet and stairway.

Nature's "sense of humor" can seem weird. In 1989 a tornado tore the roof completely off a building near Lawrence, Kansas. The unharmed family stood outside surveying their home, which was entirely intact except that it no longer had a roof. The man of the family was absolutely delighted; he wanted to replace the old roof, which was full of leaks. As he and his wife debated how they would go about this, one of the children screamed that the tornado was returning. It had actually doubled back on itself and was retracing its former path. The returning tornado settled the debate. This time around it removed the entire house.

Nothing is stranger than what occurred in summer 1938 in the town of Baird, Texas. The citizens of Baird were uneasy that day as the air felt familiar to most of them. They were sure it was tornado weather, and they were right.

Late in the afternoon an ominous black cloud appeared rather suddenly and moved rapidly over the town. It was very definitely the type of cloud that spawns a tornado. Baird residents could see the thick, feathery cloud wisps moving in and out of each other, leaving streaks of black or white whipping cloudlets. Then a huge section of the main cloud began to rotate, and a twister formed and descended rapidly on the town. As the twister, a whitish whirling fluff, hit the ground, its bottom immediately became black with churned-up dark soil cover. The twister rapidly turned black from the ground up into the main cloud, which by now was also spawning several small tornadoes that would whirl off the cloud a relatively short distance and dissipate. This activity seemed to be continuous.

The tornado that touched down was the killer. It cut a path 200 yards wide and nearly a mile long, almost completely through the town. Any house in its path was utterly destroyed; in all, almost 50 homes became matchsticks. The tornado, seemingly enjoying itself, started to return to the mother cloud. As it rose it passed over a house and tried to take the building with it. The house, lifted from its foundation and turned at least one complete rotation, dropped about 60 feet away. But not all of it was destroyed. The tornado lifted the house from its base with such sudden force that the floor stayed behind intact, with all the furniture still standing upright and in place. In the bathroom a man was busily taking a bath. Completely unaware that a tornado was in the process of destroying much of Baird, he was in the act of soaping himself down when the house suddenly disappeared from above him. And there he sat in his nakedness, completely bewildered, frightened, unharmed, and with only the sky for a ceiling.

The Hanging That Prompted a Library

The life of copper miners in old Arizona was not an easy one. Companies, more interested in profits than in the welfare of their workers, were indifferent to the deplorable mining conditions. Common mine incidents included being blown sky-high by a dynamite blast or asphyxiated by noxious gas. Cave-ins and fires were a constant menace. Faulty equipment was almost routine—for example, the cable controlling an elevator cage that broke, sending nine miners hundreds of feet to their deaths.

The advent of steam drills only made conditions worse, because the drills kicked up clouds of rock dust for the miners to breathe. The drills were called "widowmakers" in cynical recognition of the deaths that resulted from the dreaded silicosis of the lungs. The only restitution a widow could expect after an industrial accident was whatever the local citizens might collect for her. Men worked deep in the bowels of the earth on 10-hour shifts, six days a week, and earned three dollars a day. This was known locally as the era of the 10-day miner, for that was about as long as the average laborer could endure the intolerable working conditions.

In 1880 a small eastern mining firm known as Phelps-Dodge and Company sent Dr. James Douglas to Arizona to check out the prospects for a mining venture. Douglas, a physician turned geologist, sent back favorable reports on the ore potential of the region. Dr. Douglas moved to Bisbee and became an assayer for the booming mine that came to be known as the Copper Queen.

Bisbee, a town of miners, some with families, had grown around a canyon appropriately named Brewery Gulch. The canyon was lined with saloons that provided an overabundance of relaxation for the miners. The reputation of the saloons was that the farther up the gulch they were, the rougher the clientele. Beyond them were the sporting houses; all that remains today are the concrete stairs that led up to the wooden shanties where the girls worked—symbolic, perhaps, as remnants of a way of life that, for most, led nowhere.

Most of the town's houses were built on slopes so steep that, as local people would attest, anyone falling "off" the front yard would land on a neighbor's rooftop. The social status of residents was determined by how high they lived on the hillside. The most affluent part of town, where the highest-ranking company officials lived, was called Quality Hill. An important feature of the neighborhood was a cave in which company owners and managers could (and did) lock themselves away during serious labor disputes. These happened with some regularity.

Bisbee lived up to its reputation of having sunshine 330 days of the year, but moonshine was there every day. Moonshine whiskey, known locally as "bug juice," provided the inflammatory spirit that gave the town its sensational aspects. Indirectly, bug juice also provided Bisbee its most unusual resource, something lacking in all other existing Arizona towns: a library.

On an unremarkable Sunday afternoon in 1882 some miners were playing twenty-one in the Bon Ton Saloon. The usual drunken argument took place, and a man involved in the dispute shot three times, killing an innocent bystander and wounding a miner and the surprised bartender in the adjacent saloon. The miners promptly discharged their civic obligation by hanging the man from a cottonwood tree near the center of town and then returned to the saloon. They continued their game of twenty-one to determine the loser, who would be responsible for cutting down and burying the hanged man. This led to more disputes and additional shooting. Meanwhile the dead man continued to swing in the breeze.

At this moment three men came riding by: Dr. James Douglas and two gentlemen from New York—Mr. Phelps and Mr. Dodge, who had come to see their newly acquired property. The three men were horrified at the sight of the swinging corpse. They resolved that the miners needed a diversion, a means of relaxation during their time off so they could develop better spirits and character.

A new building was soon erected in Bisbee near the lower end of Brewery Gulch and was endowed as the town library. Here the men could read the works of Homer, Shakespeare, and Milton, with Chaucer, Dante, and Goethe as lighter fare. That the miners had never heard of these well-known authors didn't matter, since more than 90 percent of them could not read anyway.

During the early days of the library it served as a meeting place for the miners. They didn't read any of the books, but the card games were pretty wild. To endow the library users with some shred of refinement, no liquor was allowed on the premises of the Copper Queen Library.

The library is still there!

Then Comes the Undertaker Bee

On any normal summer day, each of the 20,000 to 60,000 workers in a colony of honeybees will, obsessively and cheerfully, go about the specific job to which it has been assigned. In the meantime, the queen is busy laying 1,500 or more eggs per day and depositing them in wax cells. She cannot take care of her young and is herself totally dependent on the workers for food and care. The eggs will hatch into larvae in three days, at which time the job of the worker bee begins in earnest. The helpless larvae must be fed approximately 1,300 times a day (54 times per hour) during their period of growth. During the six or seven days it is developing from egg to pupa, a single larva will gain up to 1,500 times its birth weight.

The task of each worker (all infertile females) is determined by its age, for it is physically equipped to handle different specific duties at different stages of maturity. The moment that an adult emerges from the pupa, its first assignment will be to clean out the brood cells and make them ready for the next batch of eggs. Within a few days the worker has developed food-secreting glands and becomes a nurse bee ready to feed the hungry larvae. Nurses pass nectar back and forth, adding essential enzymes, and they wean older larvae to a diet of pollen and honey.

By the time the worker is 12 days old, the nurse glands dry up and wax-producing glands develop. The bee is now able to build and repair the comb, make and store honey, clean, protect, and regulate the temperature of the hive. After about 10 days the wax glands cease to function and the hive workers are ready to venture outdoors to forage for nectar, pollen, water, and resin (for sealing cracks in the hive)—the only materials needed to keep the colony thriving.

In a single summer day, the foragers may collect nectar from over 250,000 flowers, making about 3,000 trips to fetch a pound of nectar. This involves a combined flight distance of at least 300,000 miles. Small wonder this is described as "the killing job of the worker bee." By the end of three weeks of foraging, at the age of 42 days of adulthood, the worker, with tattered wings and a worn-out body, will die. For this anticipated event, the hive is prepared.

In the highly efficient organization within the bee colony, there are many specific subdivisions of labor. Among the nurse bees are those specially selected to attend the queen, grooming and feeding her. The few hundred drones, who wait in a kind of stupor for some vague future nuptial flight, must also be cared for. The foragers collect all of the essentials needed by the colony, careful not to mix pollen and nectar. Those that become familiar with the neighborhood scout around for the best food sources and communicate their findings by an elaborate bee dance, which identifies distance and direction as well as how sparse or spectacular the crop. The house workers also share many divisions of labor within the hive. One of them is the undertaker bee.

Most bees will ignore the body of a dead hive mate, but within a short time after a honeybee dies, the specialized undertaker bee arrives on the scene. She will grasp the dead body and carry it about 400 feet from the hive. If the dead bee is not removed, disease-bearing vermin will quickly arrive, or, worse, the decaying body could attract predators that would decimate the community.

Researchers are not sure how the job of undertaking is assigned to specific bees, but all are two- to three-week-old workers. Although the queen produces hormones that control various aspects of bee behavior, the workers seem to decide which among them shall do the undertaking. Only a small number of bees share the responsibility for removing dead bodies during the summer season, but when autumn sets in, other work crews will join them. This is when the hive is readied for winter. There will be no place and no food for a nonproductive bee.

Only a few of the drones will have fulfilled the mission of mating with the queen (after which they die). The rest will have to go. Because autumn is a slow season for the wax-producing workers, they can join the undertaking and cleaning bees in removing useless debris from the hive. The remaining drones fit the category of useless debris, so they are turned out of the hive. They will die of cold or starvation, and the colony will continue to prosper through another season.

Catch a Falling Star

During the most recent ice age, a giant ground sloth (*Megatherium*) was peacefully feeding on a leafy tree when it noticed a light in the sky. The light appeared motionless but became brighter and larger by the second. The impossible was about to happen. The ever-growing light was a four-pound stony meteorite on a collision course with the sloth. It hit like a bullet from a high-powered rifle. Broken skeletal remains and their association with the missile from outer space leave no doubt. This event some 20,000 years ago is the earliest known instance of a living creature being injured by a meteorite.

Anyone who examines the dark sky on a clear night should see a "shooting star" every 10 to 15 minutes. These streaks of light are really sand- to pea-sized bits of comets, billions of which enter our atmosphere each day at speeds 50 times faster than a rifle bullet. Nearly all of them disintegrate into dust, but the total increases the weight of the earth by about 25 tons of interplanetary detritus each day.

Occasionally a large missile, four pounds or more, will get through the atmosphere and strike the earth; such meteorites are usually fragments from asteroids. Between the orbits of Mars and Jupiter is the asteroid belt, sort of a celestial rock garden consisting of billions of rocks that never coalesced into a planet when the solar system was being formed. These rocks are constantly colliding, and pieces break off. The force ejects fragments into another orbit that may take the extraterrestrial rock on a collision course with the earth. The ice age sloth mentioned above was a victim of an asteroid fragment.

Reports have surfaced of persons being struck by meteorites. One, a monk killed in Milan in 1650, is unsubstantiated because "a rock falling from the sky" was not an accepted phenomenon in those days. A 1954 report of a woman being struck by a 10-pound meteorite that burst through her roof, bounced off a radio, and bruised her leg has been authenticated. Considering the number of things that fall from the sky, it must seem surprising that only a few questionable injuries to living creatures can be dredged up in 20,000 years. Even a confirmed pessimist would scarcely run for cover with those odds.

An estimated 500 meteorites a year survive passage through the atmosphere and strike the earth. Throughout history, only about 2,100 separate meteorites have been validated. Recent exploration in the Antarctic has almost doubled that number. Of course, many meteorites break up at or before impact, so myriad fragments occur.

An unusual episode occurred in Wethersfield, Connecticut, on April 8, 1971. Several citizens of the town noticed a light in the sky. It didn't move but became increasingly brighter and somewhat larger. As the ice age sloth would have known, it was a meteorite about to collide with the town of Wethersfield. It struck a house, leaving a large hole in the roof, and bounced around the interior of the house, damaging some furniture but harming no one.

The incident did not appear unusual until November 8, 1982, just 11 years later. Some of the same citizens of Wethersfield noticed the same type of brilliant motionless light in the sky. This could mean only one thing: a meteorite was coming straight toward them. Having survived its fiery passage through the atmosphere, the six-pound, grapefruit-sized meteorite crashed through the roof of a house and then through the second- and first-floor ceilings. It came to rest in the dining room at the end of a trail of plaster and splinters. This impact was less than a mile from where the earlier meteorite had struck. Like most meteoric debris, the meteorite was not hot when it landed, had burned nothing, and was cool to the touch.

That two objects from space would land within a mile of each other is infinitely more incredible than the fact that so few creatures have been hit by space debris. Space is so vast, and the earth is so small a target that it is only because of the huge number of missiles that fall through the sky that the earth is hit at all.

The above incidents involved asteroid fragments of 10 pounds or less. But larger asteroids, several over 100 miles in diameter, have elliptical orbits that at times cross the earth's orbit. Scientists estimate that at least 1,500 asteroids large enough to be measured in miles have elliptical orbits and could collide with the earth some day. Knowing their orbits, scientists can predict a collision. By the time it occurs we hope to have the capability of either destroying the asteroid or deflecting its orbit to make it miss the earth. Thank goodness for a small-sized earth!

Meteorites in the 50,000-ton range appear to strike the earth every 100,000 years or so. Meteor Crater, in Arizona, is four-fifths of a mile in diameter and 575 feet deep, the result of a collision with a meteorite that fragmented upon impact 50,000 years ago. Near misses have occurred. Several asteroids, each less than a mile in diameter, earned that title in recent times. Hermes, in 1937, came within 540,000 miles of the earth, and Icarus, in 1968, passed within six million miles of our planet. Then, in March 1989, an asteroid at least a half-mile across, millions of tons of rock moving at 44,000 miles per hour, shot by the earth at less than twice the distance of the moon. This is so close that it is considered a cosmic near miss. An impact would have been equivalent to the explosion of 20,000 one-megaton hydrogen bombs. It would have left a crater 10 miles across and about a mile deep.

This asteroid, now known as 1989FC, orbits the sun once a year in a regular elliptical path, and it most certainly will be back. Eventually either it will collide with the earth or our planet, with a far superior gravitational influence, will sling it away into another orbit.

No one saw 1989FC go by. It was discovered by a scientist accidentally when he examined a series of photographs taken on March

31 that year with an 18-inch telescope at Palomar Observatory in California. Working backward in time after calculating the orbit, he realized that 1989FC had crossed earth's orbit on March 23. Had the earth been at that same point in its orbit on that day, the result would have been the aforementioned catastrophe.

The asteroid missed earth by a mere six hours. Were you feeling especially lucky on March 23, 1989?

The Day the Mountains Walked

Although Japan would seem to be the most earthquake-devastated country in the world, that tragic distinction really goes to China.

A great earthquake, magnitude 8.5 on the Richter scale, struck Gansu Province on December 16, 1920. The huge tremor took 180,000 to 200,000 human lives. Most deaths were caused by gigantic landslides that buried entire villages, damming rivers and turning valleys into instant lakes. The people still talk about the quake as *Shan tsoliao*, meaning "the mountains walked."

The nickname is apt, since the gigantic slides covered an area of over 30,000 square miles. In one section alone, at least 17 immense landslides occurred within a 20-mile semicircle. Millions upon millions of cubic miles of soil and rock cascaded into adjacent valleys and plains. The scars left by some of the slides were so clean that they appear to have been scooped out by a gigantic trawl.

The monstrous avalanche of soil and rock was the real destroyer of life and property. Some villages were blotted out entirely: men, women, children, livestock, and dwellings were buried forever under immense blankets of earth. Only handfuls of people who lived in the outskirts of inhabited areas opposite the source of the slide managed to survive.

Into one three-mile area, now rightly called Valley of the Dead, poured seven landslides, wiping out the entire population except a man and his two sons. They were inside their house located high on the mountain's side when the quake began. Their home, almost instantly caught up, rode on the crest of the rockslide for at least a half-mile. These three slides met, whirling the house and its inhabitants as if in a giant maelstrom and carrying them for another half-mile to the valley floor. Somehow the house and its inhabitants survived; although the interior of the building was a shambles, the frame was intact. When the man and his sons stepped out through the doorway, they saw they were on the valley floor and wondered where the village was. It did not take them long to realize that the entire village lay buried beneath them, a large cemetery.

In another part of the Valley of the Dead two sections of an ancient, well-packed highway including the bordering trees were swept across a stream and set completely intact on top of a mass of loose earth. In a neighboring area one colossal slide plowed deep beneath a mountain roadway and carried a quarter-mile section for a considerable distance. When it finally came to rest, the road was virtually intact, complete with bordering trees with birds' nests in their branches. Some still contained eggs, quickly claimed by the parent birds who had followed the slide almost a mile from its point of origin.

In yet another valley, a man slept through the entire earthquake. Looking out of his window in the morning, he was astonished to find a large hill almost within arm's length of his home. A relatively large neighboring hill had slid intact onto his homestead and stopped just a few feet from his house.

In one area of Gansu Province a rockslide completely buried a large village and appeared to have marked the site of the buried village symbolically. A newly constructed temple from the mountaintop was deposited, almost gently and nearly intact, on top of the village site. The building is now regarded as a holy marker, a tombstone for the citizens of the village buried beneath. Annually on the anniversary of the catastrophic event, services are held inside the repaired shrine, and priests pray for the souls who rest beneath this temple.

"Who Murdered the Veterans?"

During the Depression years, 700 World War I veterans were employed by the federal government to build roads at the Matecumbe Keys, Florida. At the nearest town to the mainland a train sat in readiness to evacuate the men in case of a hurricane. However, when the storm did occur, in September 1935, the train did not leave for hours. When it did leave, it was destroyed by the storm. The cause of the long delay in getting the rescue train moving is not known for sure. The only real conclusion was that pure neglect doomed many of the veterans.

The men were lashed by winds blowing 200 miles per hour while great storm waves swept many to sea. When the storm abated and rescue teams arrived they found total destruction; almost nothing man-made was standing. As Ernest Hemingway later wrote, "Who murdered the veterans?"

Herding Tendencies of Dinosaurs

No doubt many dinosaur groups traveled in herds. In the Bandera County fossil quarry, in Texas, the tracks of 23 individual apatosaurs all move in the same direction, with the prints of the young in the center. One or two carnivores had passed earlier, but the herd definitely crossed as a unit.

In Holyoke, Massachusetts, are the tracks of 134 three-toed bipedal dinosaurs in three trails. Browsing would have led to erratic prints as the animals moved to and fro; these were traveling as a group sharing a common objective.

The sight of these great herds of dinosaurs moving through a Mesozoic forest with the young, often the size of a full-grown elephant, protected by a ring of attentive adults must have been one of the most awe-inspiring sights of any period in earth history.

The Terrible Bird

In October 1769 Captain James Cook stood on deck as his ship, the *Endeavour*, approached New Zealand. There he is said to have observed a huge bird on the beach; it turned and ran into the woods. If this tale is true, Cook was the only European to have ever seen a live giant moa.

New Zealand, a living museum of natural wonders, was a fitting home for such an unusual bird. Its two main islands, North Island and South Island, were once part of Gondwanaland, the supercontinent of the Southern Hemisphere. But somewhere between 60 million and 70 million years ago, New Zealand broke away from the rest of the world and became two isolated islands that "drifted" to their present position. The isolation of the islands produced a rare and distinctive flora and fauna. Today three-quarters of New Zealand's indigenous plants live nowhere else on earth. Its fauna, likewise unique, include living fossils such as the tuatara, a lizardlike reptile more ancient than the dinosaurs. All native birds are flightless, including a parrot that makes its home in underground burrows like a rabbit.

Before the arrival of humans, New Zealand was free of all predators, so the birds, having no enemies, lost the use of their wings. Then the Polynesian Islanders arrived, by accident in the 10th century and

for the purposes of colonizing with the arrival of the Great Fleet about A.D. 1300. These Maori, with their dogs and (inadvertently) their rats, were the predatory mammals for whom flightless birds became easy prey. Unfortunately, wings can't suddenly develop and start flying when enemies appear.

The Dutch explorer Abel Tasman passed by the islands in 1642 and named them New Zealand. But the Maori were sole proprietors over the land of the moa until 1769, when Captain Cook landed there. The first European had arrived, and the Maori were forced to make room for white settlers who would follow. The natives described moas and moa hunts, but fact was diluted with legend. Then, in 1838 Dr. John Rule obtained a battered thighbone of a huge creature. He brought it to England to Sir Richard Owen, the leading authority on animal anatomy. Owen dismissed it as an ordinary soupbone at first but, after comparing it with an ostrich skeleton, recognized it as a thighbone of an impressively large bird. In 1839 Owen reported the bird to the Zoological Society of London and gave the moa its first official scientific recognition. He gave the moa the generic name of *Dinornis* ("terrible bird"), believing it to be at least as large as an ostrich. After several cases of moa bones were sent to him for examination, he had plenty of bones to reconstruct entire specimens.

Zoologists today classify the moa into five genera and at least 24 species. The smallest moa was about the size of a turkey, and the tallest, the 12-foot *Dinornis maximus*, which must have weighed at least 600 pounds, was twice the height of a man. Many bones and other remains identify the moa as a massive, ostrichlike bird with a long neck, sturdy legs, and hairlike feathers similar to the kiwi's. Footprints show three large toes with heavy claws spanning 14 inches and a stride of over 30 inches. No moa remains show traces of wing bones, so ancestral moas must have lost the power of flight millions of years ago. Radiometric dating of a fossilized moa bone gave an age of 35 million years, clearly demonstrating the antiquity of this genus in New Zealand.

In 1939 two scientists discovered a sticky, muddy swamp that had been a natural trap for moas. The heavy birds venturing onto the deceptive crust became mired in the clay and couldn't get free. Complete skeletons were found standing upright, buried in the quicksandlike deposits. They proved that Owen's deductions based on the single thigh fragment were quite accurate. The largest species of moa found here was about 12 feet tall and weighed close to half a ton. Carbon-14 dating showed that the swamp had developed about 1700 B.C. Another radiometric date showed that the largest moas were still being trapped in the swamp as recently as 675 years ago. By this time the birds' biggest headache had already arrived, because now they had become important food for humans.

To zoologists the most important characteristic of the giant moas was the fact that the skull of a 12-foot moa was no larger than a poodle's. Compared with its immense body the skull was insignificant, a microcephalic pinhead perched atop a long, graceful, tapering neck. The moa's adaptations are well known, for several specimens have been found with eggs, some containing unborn chicks. A few skeletons still have pieces of skin adhering to the bones, and there are also vestiges of feathers, footprints, and (most important of all) gizzard linings.

The moa had to eat continuously to supply fuel for so large a body. With such a small head and no teeth to chew with, though, it first had to solve the problem of how to grind its food down to proper size for digestion. As with its ostrich cousin, and other seed- and grain-eating birds, the moa swallowed a number of pebble-sized stones that were retained in its gizzard. They very efficiently pulverized the vegetation swallowed by the puny-headed moa. With a way of chewing internally, moas didn't need teeth.

Most of the stones associated with moa skeletons were types found nowhere else in the entombing sediment. For miles around the sites of moa remains scientists searched for the original source of country rock containing the type of stone the gizzards contained. Searchers found that the birds had often traveled over 10 miles to acquire the right pebbles. Such careful selection suggests that the moas were driven to find the hardest possible rocks for their gizzards. Undoubtedly, long, energy-expending searches were involved, implying that very special rocks were essential to process enough food to support their metabolism. Moa gizzard stones are highly polished and glisten with a fine patina achieved by almost continuous grinding of the hard pebbles against each other. Only the pulverizing of great masses of leaves and twigs could satisfy the food requirements of the moa.

For millions of years before humans arrived the moa had reproduced itself adequately, probably laying one or two eggs per season. In the absence of predators, reproduction was controlled appropriately by nature. When the Great Fleet of the Maori arrived, the moa were abundantly distributed throughout the two islands. New Zealand must have seemed a land of plenty to the ancestral Maori, with its cornucopia of juicy, flightless birds. But the moas' slow rate of reproduction helped cause their extinction, because they couldn't keep up with the human demand for eggs and birds as a basic food supply. Between A.D. 1300 and 1800 they were exterminated by the Maori. Probably most were gone by 1500, with only a few remote stragglers lingering into the next century or two.

All Maori archaeological sites show an abundance of moa bones. The oldest sites contain the greatest profusion of moa eggs and remains. Scientists know that the natives ate the flesh and eggs, adorned their hair with the feathers, made fishhooks from bones, and crushed the skulls and tattooed themselves with the powder. Mounds and cooking pits often contained the remnants of a huge moa feast. Their importance to the Maori was revealed in the discovery of a tomb containing a human skeleton in a sitting position. In the skeleton's hand was an entire *Dinornis* egg, 10 inches long, apparently placed there for nourishment during the journey to the next world.

Early white settlers described the method of hunting. Hunters would surround the moa, which when attacked would stand on one leg and kick with the other. Several hunters would keep the moa's attention while another sneaked up from behind with a heavy club and hit the leg that supported the bird. As it went down they would finish it with spears. The hunts often had unplanned results. At times the agile moa, sensing the man sneaking up on it from behind, would turn on the skulking human with a powerful kick that quickly ended the hunter's earthly problems. Sometimes, instead of kicking, the bird towering over the hunter would come down with a powerful swipe of its neck, toppling the hunter. When a death occurred the surviving hunters pursued the killer bird with a vengeance and never let up until the moa was captured. The family of the dead man was usually awarded the choice pieces of the bird during the meal that followed.

After 1500, as the moa became increasingly scarce the Maori began to look around for other forms of meat. The only animals large enough to feed them were other human beings. Cannibalism became the logical solution. Once it began, a permanent state of war existed between all the tribes. Archeological sites from A.D. 1500 to 1800 showed fewer and fewer moa bones and more and more human ones. Warfare became a sacred institution, often followed by ritual feasting: leading warriors consumed the heart, and the hunters gained courage by drinking blood. Cannibalism continued almost completely unchecked well into the 19th century.

Smaller species of moa may have survived into the mid-19th century. Stories of moa hunts were still told, but information was from recollected memories. One of the last was told in 1868 to Sir George Grey, the governor of New Zealand, when he met a party of Maori hunters. They enthusiastically told him of their recent successful hunt, the killing of a small moa. According to the governor, "they described it with so much spirit" he believed they were telling the truth. If so, they may have had the dubious honor of killing the last of the moas.

CHAPTER FOUR

Petrified Lightning

During the few seconds it takes to read this paragraph, lightning will strike the earth about 700 times. This rough estimate of lightning discharges at the rate of 100 times per second represents, worldwide, over four billion kilowatts of continuous power! Lightning is one of nature's most powerful forces, causing incalculable damage and, quite frequently, death.

Utility companies worry about the intense surges of lightning-induced current that play havoc with power and telephone lines. The airlines worry about planes in the air, which essentially become flying lightning rods; commercial jets are struck by lightning on the average of once a year.

Because the extremities of aircraft, such as wingtips, nose, and tail, are usually the targets for lightning, most of these strikes do very little harm. Of course, exceptions occur, as in 1963 when a bolt struck a fuel tank of a Boeing 707 in a holding pattern near Elkton, Maryland. The explosion that followed brought down the plane, killing 81 people.

NASA, extremely wary of lightning, continues to speculate over how to avoid strikes to space vehicles without calling off a launch every time

a cloud is in sight. Their fears are well founded. In 1969, as Apollo 12 rose through what appeared to be a fair-weather cloud, it triggered two successive lightning strikes. Fortunately they caused only nondisabling damage to the outside of the spacecraft and some computer scrambling, so the astronauts continued their trip to the moon. Trouble occurred again in March 1987, when a lightning flash triggered the launching of an unmanned Atlas–Centaur rocket that broke apart under the sudden stress. A few months later NASA was again victimized when a lightning bolt triggered a premature launch of three small rockets; they plummeted unceremoniously into the Atlantic. Ironically one of these rockets was designed to study lightning.

The chances of getting hit by a "bolt from the blue" are proverbially small, yet deaths from lightning strikes average 85 persons per year. This is higher than the death toll from any other act of nature except flash floods. A direct hit by lightning packs enough electrical punch to fry skin and even explode organs, so the unfortunates in its path have a high rate of fatalities. Of course a lightning strike does not wipe out villages, and each victim is usually a single target.

When Benjamin Franklin flew his kite into a thunderstorm in 1752, he was exceptionally lucky to avoid being killed. He managed to draw a lightning bolt down through the kite line and into a key and perceived an electrical spark as he touched his knuckle to the key. Doubtless thrilled with his discovery, he remained unaware that he should have been doubly delighted at having lived through the experiment. A Swedish physicist trying to repeat Franklin's experiment a year later with a lightning rod instead of a kite was killed instantly.

Even nondestructive lightning often leaves a signature. In July 1960 two field geologists were conducting a geological survey in southeastern Arizona when a typical summer thunderstorm descended on them. The rain was so torrential that they could see only a few feet in front of where they stood. The geologists decided to get out of the desert (which was rapidly becoming a lake); their workday was obviously at an end. As they headed for their jeep, one turned to close the gate so the cattle wouldn't get out, and a very loud explosion completely enveloped him. The ground suddenly came up at him, and his feet felt as though they were carrying hundreds of pounds of weight. He went to his knees.

As the stricken man regained his posture and composure, his alarmed partner hurried toward him asking if he was able to breathe. Relieved when the answer was "Yes, why?" the partner explained that a near lightning strike usually renders the victim unable to breathe. He was sure that the bolt had struck the gate-closing geologist—who, typical of people struck by lightning, never saw the bolt.

The day after the deluge the geologists returned to the scene of their encounter with lightning. At the precise spot that the bolt had struck, about two feet from where the man had stood, they found a greenish-brown, glassy, encrusted tube sticking out of the ground. One of the geologists remarked, "Looks like we have a fulgurite!"

When a lightning bolt strikes a loose surface of unconsolidated sand, it tends to fuse the sand (which melts at 1,710 degrees Celsius) into a glassy tube of silica, usually about half an inch thick. The tube may extend several feet beneath the surface. The largest ever found, in

Cumberland, England, had a diameter of two and one-half inches and was over 40 feet long.

As the tube goes downward it tends to branch, giving rise to numerous tubelets. It may contain fine glassy splinters on the surface and is often delicate to the touch. The splintery surface of the tube may be an image of the electrical lightning bolt that created this mineralogical wonder. That's why fulgurites are often referred to as "petrified lightning."

Shark Pen

"Fear is a motive for establishing a god and man has worshiped at one time or another most of the living things he fears."

"Where Sharks Are Gods"

Sharks play an important part in the lives of many of the island peoples of the Pacific. Shark gods are still worshiped in some island communities, although once-common human sacrifices are no longer offered to these fearsome gods. This reverence toward the shark confirms the truism that fear is a motive for establishing a god. At one time or another, humans have worshiped most of the things that they fear.

In many primitive religions shark worship grew into a complex belief, apparently attempting to reconcile the shark's roles as a cunning devil and vengeful god. The occasional human snatched from the sea could scarcely satisfy this awesome deity, so the shark gods demanded the ultimate homage: human sacrifice. In some Pacific islands the high priest, at a fateful time, went among crowds of people carrying a noose similar to a shark snare. He would arbitrarily hurl the noose into the crowd, and the man, woman, or child upon whom it landed was ritualistically cut into pieces and flung into the sea for the ravenous shark gods.

Tales of sharks are an important thread in the rich tapestry of Hawaiian legends. From the time that myths shrouded the islands, storytellers would begin, "I will tell you of Kamo-hoa-lii, the king of all the sharks." Many stories that were told by fathers' fathers, including tales of Kamo-hoa-lii, are still told in various parts of the islands.

Hawaiian reverence for the shark contrasted sharply with the attitude of early whalers and other seafarers of the western world. When a shark was caught and hoisted on board, the sailors seemed to imitate a shark's feeding frenzy as they ritually mutilated it in an orgy of blood and fury. That the shark embodied all that was evil seemed to justify the purposeless and shameful slaughter.

During the early 1900s, when the U.S. Navy started to build a major sea base at Pearl Harbor, the dredging operations unwittingly destroyed the remnants of an ancient shark pen. It was immediately identified by all the native laborers present. Here, ages before, Hawaiian kings had hurled living humans to the royal sharks, and contests were staged between ravenously hungry sharks and native gladiators.

Halfway around the world, Roman gladiators were entering an arena to battle lions to the death for the entertainment of the imperial rulers and aristocrats. Here Hawaiian warriors (coerced "volunteers") entered the shark pen and engaged in death duels with sharks. The only weapon used by the Polynesian warrior was a shark-toothed dagger, a wooden rod less than a foot in length and shaped like a stout broomstick, that he gripped firmly in his hand. Protruding from the stick were several shark teeth for cutting and a point for stabbing. The warrior had a single life-or-death chance. As the shark charged him, at the last instant he had to dive under the shark and rip open its belly with his crude knife.

Hawaiian legends tell of warriors who were successful in killing the shark, but this must have been very rare. Royal edicts stipulated that the warrior need only draw blood and then he could leave the pen. He would, of course, have to escape a slightly wounded shark, so the duel between man and shark could seldom have ended any way other than for the man to become the shark's dinner.

A shark pen was built with a large circle of basaltic rocks enclosing about a four-acre area at the edge of a harbor. The circle had an opening on the seaward side so that water would flow in. Fish and human bait were thrown into the pen to lure sharks through the open passage. During a contest the pen opening was sealed so neither the shark nor its adversary could escape.

Close to the shark pen at the bottom of Pearl Harbor lived the Queen Shark, according to legend, guarded by two stalwart sharks from each of the Hawaiian Islands. She permitted the gladiatorial contests near her royal lair as long as she received offerings—these being, of course, human dinners. In old Hawaii, humans were economically of less value than pigs.

Part of the harbor project mentioned earlier, in which the ancient shark pen had been destroyed, was the construction of a huge four-million-dollar dry dock. For unknown reasons the foundation collapsed, and the dry dock was destroyed. Naval engineers and construction workers scurried around trying to find the cause, but the native laborers knew what had happened, and why. They stated simply, "Queen Shark is hubu (angry) and humps her back because she hasn't received any offerings."

Prehistoric Pit Barbecue

In the cave of Teshik Tash, Siberia, evidence abounds that its Neanderthal inhabitants specialized in hunting the Siberian ibex. The Stone Age Neanderthals were not noted for keeping a tidy cave, so bones of numerous ibex were strewn throughout the living area. The cave may have been purposely unkempt. Scraps of food would attract small animals, who would become a supplementary treat for the tenants. Numerous ibex remains indicate that they were a favorite prey, although their size and menacing horns suggest that they could not have been captured without a struggle.

Deep, narrow ravines, a prominent feature of the landscape throughout the region, provide a clue to the method the Neanderthal used to snare the ibex. With raucous shouting and waving of spears and clubs, one group of hunters would stampede the herd over a cliff. The ibex would tumble into the ravine, where they lay with broken bones, to be easily dispatched by hunters waiting below. The bodies were then carried to nearby caves and eaten. Radiometric dating shows that these ibex met their fate over 65,000 years ago.

In the last half-century the outdoor barbecue has become an essential fixture in suburban American lifestyle. Occasionally in the interest of glamour and a convivial group effort, the expensive, well-equipped grill is replaced with a simple firepit. When rocks on the bottom are heated sufficiently, some unfortunate animal, usually a pig, lamb, or any kind of fowl, is placed in the pit and roasted to provide a "new" delicious high in outdoor living. However, the pit barbeque is actually a "hand-me-down" from one our prehistoric ancestors used for thousands of years.

Scientists investigating early inhabited sites in Europe have often been impressed by the vast numbers of large game animals hunted and consumed by Stone Age citizens. These included such behemoths as the mammoth, mastodon, giant sloth, cave bear, and woolly rhinoceros. The question uppermost in the minds of the researchers was how early man could, with his primitive tools and weapons, subdue such many-tonned beasts and drag them home to dinner.

Throughout much of France and Spain remnants of deep pits are rather common. Here, scientists speculated, lay the answer. The ancient hunters would dig pits and, with shouts, pounding, and whatever racket they could kick up, would drive the animal toward the pits. Some would blindly fall into the disguised hole in the ground. In a pit a huge animal was practically helpless, and the hunters would bounce rocks off its head or spear the animal without any real danger to themselves. Then they would cut up the meat and transport sizable portions, or perhaps leave the animal in the pit and feast on it there.

No pit was ever found that contained the animal itself, so this logical scenario was unproved until quite recently. At the base of a 150-foot cliff near Badegoule, France, scientists discovered the bones of a 12-foot, eight-ton ice age mammoth. The remains indicate it was the victim of a successful hunt, probably by a group of Cro-Magnon humans. The mammoth appears to have been a solitary grazer near the edge of the steep cliff. The shrieking hunters, waving spears, clubs, and fire brands, would have driven the huge beast into a panic until it turned and ran headlong off the cliff. The impact of the fall broke all four legs and probably killed the animal outright. In a deep pit dug under the carcass, the mammoth was roasted right on the spot, undoubtedly amid great celebration. Radiometric dating indicates the hunt happened about 30,000 years ago.

A large ice age animal was found in another pit, not in Europe but in the Patagonian region of Argentina. During the early 1980s scientists uncovered the remains of a deep pit into which a giant ice age sloth had been driven 12,000 years ago. After driving it into the pit, the Paleo-Indians quickly dispatched the huge animal. They never attempted to remove it but rather cooked the animal right where it fell. It probably weighed several tons and certainly provided enough to feed the tribe of hunters.

The efficient cooking style of choice for many thousands of years was the pit barbecue. Only recently have glamour and elegance been added to this cooking method.

The Great Mouse War

On November 24, 1926, as rain began to fall on Taft, California, an offbeat chain of events was set in motion, the strangest occurrence of its kind in the United States.

About seven miles from the city was a dry lake bed called Buena Vista Lake, in which farmers had planted over 11,000 acres of grain. This lake bed, implanted with so much grain and seed, had became home to millions of field mice. As the rain came down in earnest the dry lake bed began to fill with water. Somehow a general distress signal went out among the mice, and they began a sudden mass exodus from the lake bed—millions of mice forming a huge undulating ground cover.

Quiet Taft, an oil and agricultural town located in the desert of the southern San Joaquin Valley, was unprepared for what was about to happen. From out of the barren landscape millions of mice descended on Taft in search of food and shelter. The oil town was besieged by a living carpet of fur.

The advancing mouse army met their first real resistance when they encountered humans at the Honolulu Oil Lease, located about three miles from Buena Vista Lake. At first, the mice were a big joke to the oil workers. But when the men became completely surrounded by almost solid balls of fur, with small teeth nipping at their pants legs and ankles and occasionally drawing blood, the joke was over. The men began to swat at them with shovels. This caused many mouse casualties, but it was like spooning sand from a beach. So the humans beat a rapid retreat and quickly organized a plan of warfare against the vast army of mice.

They plowed a series of trenches a foot or so deep and sowed them with poisoned grain. This tactic was so successful, the men crisscrossed the countryside for miles around with shallow trenches baited with lethal grain. Even local government bodies set up command posts and dispensed poisoned grain to the residents. The mice died by the millions but were immediately replaced by millions more. After about two weeks of intensive assault, an editorial appeared in Taft's newspaper, the *Daily Midway Driller*, pleading for citizens to "page the Pied Piper." Cats became the pied pipers of the day, while the piper of literature and legend remained in obscurity.

Despite all the casualties, the little rodents persisted and, with new generations joining them every month or so, continued to grow in numbers. Millions of rodents—as many as 100 million—dominated the landscape.

Considering that the population of Taft at that time was under 5,000, humans were outnumbered about 20,000 to 1.

The rodents' front line was probably pushed in an ever-widening circle by the pressure of the multitudes behind them. By January they had invaded Maricopa, 14 miles from the lake. On the San Emidio Ranch, eight miles away, starving multitudes of mice overcame a penned-up sheep and quickly consumed it; the implication of famished grain-eating mice becoming carnivorous was that humans could also be vulnerable. The northern extension of the hordes swarmed over the Taft–Bakersfield Highway where untold thousands became road casualties. But the mice continued to run unchecked over the land until January 22, when the "Pied Piper" arrived.

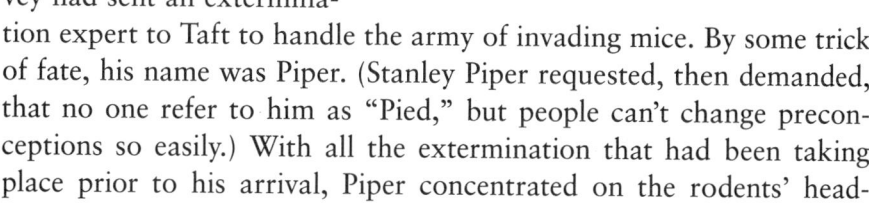

The U.S. Biological Survey had sent an extermination expert to Taft to handle the army of invading mice. By some trick of fate, his name was Piper. (Stanley Piper requested, then demanded, that no one refer to him as "Pied," but people can't change preconceptions so easily.) With all the extermination that had been taking place prior to his arrival, Piper concentrated on the rodents' headquarters, the once-again dried-up lake bed. He estimated that 45 million to 50 million mice were holed up in that one area and immediately planned his campaign to combat the mouse army.

Before Piper could put his plan into action, effective help arrived from a completely unexpected source. It came literally from heaven. The open sky began to rain hawks, gulls, herons, owls, roadrunners, ravens, shrikes, eagles—all of which dived and attacked the mouse legions in the lake bed. When the birds left, it was because they could eat no more. They ambled or flew feebly away from the gorging field after their distressingly heavy meal. Some of the mice survived but later died of disease or were drowned in a sudden outburst of heavy rain in mid-February that filled the lake bed once again.

As quickly as it had begun, the Great Mouse War of 1926–27 ended.

Life, Life Everywhere

Life in its most basic state is the bacterium. Bacteria are among the oldest, simplest, and most common forms of life on the planet. Currently scientists can separate and identify over 2,500 different kinds, but thousands more doubtless remain to be discovered, described, and classified. Sizewise, individual bacteria are the tiniest of free-living cells, so small that one drop of water may contain as many as a million individual bacteria. And they still have enough space to move about quite freely. Imagine, for example, that a single bacterium were seated in the middle of the period at the end of this sentence. To the lonely bacterium perched in its center, the period would appear the size of an average football stadium. Generally speaking, though, bacteria do not spend much time by themselves. They are social creatures that live in large groups and seem to make up for their diminutive size by a birth rate that is almost impossible to comprehend.

Imagine that the solitary bacterium sitting alone in the middle of the period is adopted as a pet. Its owners treat it well, give it all the food it can eat and all the space it will need, and keep all dangerous bacteria-killers away. Living in a bacterial Garden of Eden, free to be fruitful and multiply without limit, it shares its idyllic garden with all of its offspring. In just three days, this single bacterium—by way of cell division approximately every 20 minutes—will give rise to an extended actively reproducing bacterial family that takes up more space than does the planet earth. In those three days it will have created its own planet larger than Earth, composed entirely of bacteria. Fortunately, life is full of hazards for bacteria, and untold hordes of plants and animals consume them by the millions.

Being one of the earth's oldest inhabitants, reproducing in the simplest way (by fission, in which a cell grows until it breaks in half, creating two individuals), bacteria have evolved into many forms and have adapted to highly varied environments. This simply means that bacteria are virtually everywhere. The bacteria that perform a useful function, such as those that aid digestion by breaking down food, are joined by the harmless stowaways that live on and within the human body. The most notorious of all microorganisms are those formidable adversaries capable of causing severe, even deadly, diseases.

The human mouth is a rich breeding ground teeming with many types of bacteria. On the tongue alone several hundred different species of bacteria grow side by side. Tooth decay, caused by a waste product of bacterial digestion, is a sample of bacterial handiwork.

Generally humans don't tend to give much thought to bacteria, since they can't be seen, heard, smelled, tasted, or felt. One can suddenly be reminded of them by pus-filled infections, decaying vegetables, smelly socks, or things equally repugnant. Although such signs of bacterial activity result in rather negative feelings about them, bacteria are the life-form on which the vast majority of all other living things depend.

For more than two billion years, bacteria probably shared the primeval planet with no life-form other than blue-green algae. During this time, the Precambrian Eras, neither the atmosphere nor the soil of the planet was capable of supporting complex life-forms. By the end of that period, about 600 million years ago, and largely as a result of the ceaseless work of the bacteria, the biosphere of the planet was ready to play host to a mind-boggling diversity of life. Eventually that led to relatively complex life-forms such as humans.

Among microscopic organisms, bacteria's most efficient competitor for living space is the fungus. As far as scientists can determine, fungi have inhabited the earth since early Paleozoic times, about 400 million years ago. By that time bacteria had already had a head start of over a billion and a half years. Being a lot smaller, bacteria are considered much more abundant, but fungi are also typically everywhere. A wheelbarrow full of soil, for example, contains more fungi than there are people on the earth. (That many bacteria could be found swimming in a mere 5,000 drops of water—less than $2/3$ cup.)

Throughout the eons of their existence, fungi have been one of bacteria's most competent natural prey organisms. One fungus, for example, releases a powerful chemical quite deadly to certain bacteria. Fungi that have developed this type of defense have become allies to humans in the war against infection and disease.

In 1928, an English scientist named Alexander Fleming was growing a dish of bacteria in a London laboratory. He noticed it had been contaminated by a rogue bit of fungus that had somehow managed to kill off all the bacteria surrounding it. As he observed it, the action of the fungus became even more interesting. In the long run Fleming found that the fungus was giving off some kind of liquid that was very definitely a "bacteria killer."

This newly discovered chemical substance, subsequently defined and reproduced in the laboratory, came from the penicillin family. Fleming named the liquid "penicillin." This single substance changed the course of treatment of infectious diseases and has saved more human lives than any other drug in the history of medicine. Although an English scientist discovered it, and in the 11th century the Mayan people recognized the therapeutic qualities of the mold that grew on tortillas, penicillin was actually "invented" millions of years ago by an anonymous mutant fungus.

Since Fleming's discovery numerous other antibiotics, such as amoxicillin, erythromycin, streptomycin, and tetracycline, have been discovered in much the same way—as naturally occurring fungal and bacterial products. Medical researchers believe that they have scarcely scratched the surface of the antibiotic heap. Considering that the vast majority of bacterial and fungal types are as yet undiscovered, cures to other ailments probably have already been created by some unknown fungi. They are just waiting to be discovered, possibly by the reader!

Cloak of Darkness

Television frequently offers documentaries on African wildlife. When prey animals such as the antelope are featured, one can clearly see that caution, vigilance, and the ability to make a hasty retreat are basic to the life and well-being of these creatures. Their eyes, wide open, are constantly darting in every direction; their large ears move continuously—forward, to the side, to the rear—straining for the slightest sound of an approaching predator. This watchfulness is constant; even as they feed, antelope will take a quick bite or nibble on vegetation and then resume the alert scan of the territory while chewing. Still, scarcely a day passes without one or more of their number being pulled down by some carnivore or other.

Many predators, including lions, hyenas, and leopards, are nocturnal and therefore do most of their hunting after darkness falls. As the shadows lengthen toward evening, the animal's alertness increases dramatically; it knows the hunter is hidden under the cloak of darkness. So, relying on its well-developed sense of hearing, the antelope never sleeps more than two or three minutes at a time.

With the approach of daylight, a wave of temporary relief must flow through prey animals. Shadows no longer hide the threat of death, and they may now dare to lean over a water hole for a drink.

Is it any wonder that primitive humans worshiped the sun!

The Little Ice Age

"If Winter comes, can Spring be far behind?"
                                                   Percy Bysshe Shelley

The classic Neanderthals of 100,000 years ago were the first humans to live under Arctic conditions. These early hunters had to seek out a cave sanctuary for the winter, which often meant driving out the cave bear, hyena, or an even larger, more uncooperative beast. Secure in a cave, they lived out the harsh winter.

Large pole holes at the entrance to several caves indicate that these early humans must have draped poles with large skins to block the entrance. The center of the cave was probably carpeted with warm furs surrounding a central fire. In the evenings, as the wind howled outside, the tribe huddled together on the mattress of furs, snug around a central fire. They remained warm and comfortable throughout the night, with someone assigned to keep the fire going. Artifacts left behind by the succeeding Cro-Magnon show humans clad in parkas not unlike those of present-day Inuit.

A popular misconception is that the entire globe was frigid during the last ice age. The subtropics and tropics were far less severely affected than the land to the north, but temperatures experienced in the subtropics were about 8 degrees Fahrenheit below the present average. The equatorial belt of rain forests 18,000 years ago was probably just a little cooler than it is today. Possibly many of the northern humans migrated south during the harsh winter months.

The most recent major ice age, preceded and followed by warmer times (interglacials), began about 70,000 years ago. It reached a bitter extreme some 50,000 years later and then began a slow retreat. The retreat was short lived because the glacial ice advanced again, bringing yet colder weather that reached its climax about 18,000 years ago. During this latest ice age huge sheets of ice, often 10,000 feet thick, covered much of the temperate and upper latitudes around the globe. The woolly mammoth, woolly rhinoceros, and other fauna especially adapted to glacial climates were abundant. The ice began to retreat about 15,000 years ago, slowly at first; then a little over 10,000 years before the present (B.P.), the melting phase accelerated.

In an extended warm period 5,000 to 7,000 years ago, average temperatures became significantly higher than those at present. This undoubtedly resulted in a rise of sea level and large-scale flooding of the land as some of the polar ice melted. An interesting corollary to the rising waters on earth is the appearance in the folklore of many ancient civilizations of a story of a vast flooding such as the biblical flood of Noah. An analysis of several drill cores taken from the Gulf of Mexico indicates a sudden surge in climate and a most dramatic rise in sea level caused by disintegration of one of the polar ice caps several thousand years ago. Paleoclimatologists speculate that the flooding of low-lying coastal areas, many inhabited by humans, gave rise to the deluge stories common to many traditions worldwide. Many North and South American Indian tribes, the Hindu people of India, Baby-

Ionians, and Greeks, as well as the Hebrews, are among the civilizations that included a story of a flood among their folk legends.

Around 5,000 years B.P. the trend began to reverse, with cooler temperatures prevailing worldwide. The polar ice stopped its melting trend in a climatic phase that lasted, interspersed by many fluctuations, until about A.D. 400.

The years between A.D. 400 and 1200 were characterized by warmer temperatures worldwide and fewer storms in the Atlantic Ocean and North Sea. Spain was almost tropical; forests grew higher on the mountains, and farms spread further up the hillsides. The warm conditions invited colonization by such people as the Norsemen, and around A.D. 985 about 300 people with their livestock sailed from Iceland to Greenland. They went to the grassy shores that Eric the Red had selected for them in an earlier voyage. Many generations lived in permanent colonies and prospered in a pleasant, productive climate. The Greenland colonists and their European counterparts had regular trade and communication for several centuries. Europe was their only source of iron for tools and weapons and of wood for building ships.

By about A.D. 1200 the balmy weather began to wane, and the climate again started to cool. Slowly but surely the colonies in Greenland declined and eventually disappeared. Graves excavated by Danish archaeologists in 1921 tell the story of the gradual demise of the Norse colonies in Greenland. Much evidence was unearthed. The excellent preservation of both wooden objects and clothing confirmed that the ground had been frozen during most of the 500 to 800 years the bodies had been interred. The oldest burials were the deepest, and successive graves were more and more shallow as the permafrost zone crept nearer to the surface.

Clothing of current European styles found in successive burials indicated that trade had continued for many generations. In the 13th century the expanding drift ice in the sea made the route too hazardous to navigate. Eventually the wood supply became too scarce to be used for coffins, so the most recent graves contained bodies merely wrapped in shrouds.

Any Norse settlers who survived the rigors of the changing climate and isolation almost certainly would have merged with the Inuit by the 16th century as they followed the walrus and seal southward to less icy seas. Nevertheless their colonies lasted almost 500 years. Compare this with settlements of our European ancestors, whose first colony at Jamestown, in 1607, was established less than 400 years ago.

Although the weather conditions fluctuated with a wide range of warmings and coolings, a trend toward almost constant cooling was making itself felt by the year 1550. It lasted until about 1850. This glacial epoch, now known as the "Little Ice Age," brought the greatest extensions of ice on land and sea since the ice age of 18,000 years B.P. Valley glaciers grew appreciably, and in many mountains of the world, glaciers re-formed where none had been since the close of the big ice age. People during that cold period who studied old maps were amazed to see that, in places where a glacier had lain for as long as anyone could remember, mapmakers earlier than the 13th century had drawn farms, orchards, buildings, and roads. Obviously the growing glacier had driven away the inhabitants, ingesting structures and other cultural features, and in essence had taken over the land.

Much U.S. colonial history occurred during the Little Ice Age. Colonists in New England endured winters far more severe than any of today. (Temperature readings unfortunately are absent from early colonial weather accounts because thermometers were new instruments and naturally in short supply. A thermometer was used for the first time in America in 1717.) The winter of 1740–41 hit the American colonies especially hard. Drifting snow closed roads in New England and the Middle Colonies; rivers, inland waterways, and even saltwater channels froze over.

The storm of January 1772 became known as the Washington and Jefferson Snowstorm. George Washington watched it from his Mount Vernon home and recorded in his diary that snow everywhere was up to the breast of a tall horse, and that people were shut up for 10 or 12 days "by the deepest snow which I suppose the oldest living ever remembers to have seen in this country." At about the same time Thomas Jefferson and his bride were returning in this storm from their honeymoon. They had to abandon their carriage and ride horseback over a mountain trail to reach his home at Monticello. Fortunately for the future United States, they made it!

Conditions were unusually rough during the American Revolutionary War. The legendary winter during which Washington's troops were bivouacked at Valley Forge was actually regarded by contemporary observers as "notably mild." In fact, many scientists agree that, if Washington had camped at Valley Forge just two years later than the winter of 1777–78, the sufferings of his troops would have been unbearable, and a lost Revolution would have been the probable outcome. Even by Little Ice Age standards the winter of 1780 was "the most difficult winter ever known by any person living." Washington's troops were in the New Jersey hills watching the British stationed in New York when the cold came. It froze the water between them. Sleighs and pedestrians went from Staten Island to Manhattan and crossed the Narrows to Brooklyn, and large cannons were pulled across the harbor ice. Deserting Hessian soldiers and others crossed from Long Island over the sound to Connecticut. By January every port along the North Atlantic coast was shut. Most roads were closed too. All colonies, even Georgia, were affected by that winter's cold, and from interior posts came reports of deep snow, severe cold, and much suffering from Detroit to New Orleans. Relief finally came in about A.D. 1800, when the Little Ice Age bottomed out and the temperatures began to fluctuate upward.

Will another ice age advance as temperatures cool, followed by a glacial melt and subsiding shorelines as more water is released and the earth warms? The cycles of warm and cold are far from over; there will continue to be times of brief ice advances and retreats. Since the time 18,000 years B.P. that the latest major ice age reached its peak, the temperature has slowly and periodically warmed. Between the warm periods have been at least three cycles when temperatures cooled below the global average of 59 degrees Fahrenheit (average of high and low temperatures at all latitudes and all seasons).

At present most of the world's glaciers are retreating. Readers are not encouraged, however, to hope for a return to the good old mild Mesozoic days, when the sun shone brightly and the land masses were smaller. As glaciers and ice caps melt away, the oceans will rise and the coastal regions of the continents will be submerged. The shrinking land masses cannot be expected to accommodate the exploding human population. An equally alarming effect, when the melted ice unites with our polluted oceans, will be the loss of three-fourths of our fresh water supply. There will be "Water, water everywhere, nor any drop to drink."

There is no explanation, and no shortage of theories, as to what causes major ice ages such as the four Pleistocene glacial advances from 2.5 million to 18,000 years ago. Very likely the causes are multiple, and the rarity of ice ages suggests an unusual chance combination of several factors. Several phenomena may be involved, such as eccentricity of the earth's orbit and variations in tilt and wobble of the earth's axis. It would be an incredibly rare event for them to be synchronized so that their effects could reinforce each other, but it does happen.

The earth appears now to be in an interglacial stage. For the climate to develop into a period of major global refrigeration would require much time, possibly as much as 50,000 years. Readers should feel reassured to know that they will not be here when it happens.

Pie-Eyed Elephants and Other "Substance Abusers"

Addiction to controlled substances is rare among nonhumans, but it does occasionally occur. Birds, such as the sapsucker, that subsist on the sap of trees will often find themselves too drunk to fly right, and fruit-eating birds such as thrushes have also been observed staggering about in a stuporous condition. This happens when the bird imbibes plant juice that has fermented, or feasts on overripe fruit and berries. Birds who fly under the influence frequently wind up as a meal for a hungry, cold-sober predator.

Large mammals of Africa have been filmed after they gorged themselves on overripe fallen fruits. Antelopes observed lurching and zigzagging, bleary-eyed, show all the familiar signs of being drunk. A tipsy ostrich is a sight to behold, crossing its legs as it walks, stumbling and tottering, with its long neck wobbling after a meal of alcoholic fruits. These drunken sprees are not actually addictive, even though the participants in such wildlife bacchanalia return to the place where they found the liquored-up food until it's all gone.

Catnip has often been considered an intoxicant or stimulant for cats. Cats are fascinated not by the taste but by the odor, so catnip toys are as effective as the plant itself. A mild nerve stimulant, catnip produces a "high" feeling and does little more than stimulate valuable exercise for lazy and bored cats. Rather than being habit forming, it serves as a vitamin pill for cats that engage in nothing more strenuous than changing positions during naps.

Less harmless is the locoweed, a weedy legume of the bean family native to western North America. It causes a type of poisoning called locoism in livestock animals, including horses, cattle, sheep, and goats. Although the animal must consume large quantities of the plant to be poisoned, this happens easily because once the animal starts eating locoweed it develops a preference for the plant. Signs of poisoning are loss of weight, irregular gait, loss of muscular control and sense of direction, and violent reactions when disturbed. There being no effective treatment for locoweed poisoning, killing off locoweeds or keeping animals off ranges where locoweeds occur are the only preventives of this addictive, usually fatal, condition.

Elephants are not immune to intoxication from overripe fruits, and a tipsy elephant is a sight not soon forgotten. In India the wild elephants have discovered a substitute for seasonal fruit binges: beer. Residents of small villages find that turning their crops into beer is financially quite rewarding. The problem, of course, is that elephants from the nearby forests pay regular visits in search of brewing barrels of beer. The owner of the brew wisely allows them to drink a barrel or two without interfering. But in 1991, the elephants went too far.

Beer was mass-produced in a small village in central India, where the residents had become quite proficient in beer-making and brewing was the livelihood for many families. In spring 1991 a fair was scheduled in a nearby city where the villagers planned to market their product. The smell of the fermenting beer was too much for a passing herd of elephants. They stormed the village, knocking over anything in their paths as their nostrils led them to the many barrels of beer. By the time they were finished they had drunk every barrel of beer in sight. They were quite intoxicated, and the villagers were able to drive them off with torches.

For the remainder of the night the unsteady trumpeting of the elephants echoed through the forest as the pixilated herd acted out their binge. They slept off their elephant-sized hangovers the next day. Later the village headman related, "You don't know what trouble is until you've been surrounded by 25 drunken elephants." The elephants remembered only the good time they had drinking and returned in less than two weeks for a repeat performance; they were quite perturbed that no beer was to be found. The villagers, safely hidden, were much relieved to see them stagger off; not surprisingly, beer brewing is no longer the occupation of choice in this village.

Near the Queen Elizabeth Reserve in Uganda, Africa, a group of natives had finished making several barrels of beer in preparation for a festival the next day. Just as they were storing the brew, four old elephants grazing nearby got a whiff of their favorite beverage. They decided to start the celebration immediately and charged onto the scene. The brewmasters fled as the elephants began to tank up. After the barrels were empty, the elephants went on a drunken rampage, chasing everyone in sight, overturning cars, uprooting gardens, and tearing down native houses. They finally retreated into the jungle, but throughout the night and into the next day their muffled groans and trumpeting rang through the village. Their hangovers were some comfort to the villagers, who had canceled their own celebration.

Riches from the Sea

The amount of water on the earth, an estimated 326 million cubic miles, is too great to comprehend. An enormous proportion, approximately 97.2 percent, fills the oceans of the world.

Scientists pay a great deal of attention to the exchanges of water among the oceans, the atmosphere, and the continents. This unending circulation of the earth's water supply, named "the hydraulic cycle," is somewhat complex in its extent and process. Stated simply, water is constantly evaporated from the oceans and into the atmosphere. Prevailing winds blow this moisture-laden air over the land, where the water is precipitated out as rain or snow. This water is then carried back to the sea by rivers and underground flow. The cycle, completed and repeated endless times, has persisted since oceans and lands first became the dominant features of the evolving earth.

On its trip back to the sea, flowing water comes in contact with most elements present in the upper part of the earth's crust—over half of the 103 known elements. Some of the material is dissolved and carried to the sea in solution. As a result, the mineral richness of the sea is immense; it would be a rare phenomenon for elements and minerals that occur on land to be absent from the oceans—and that includes gold!

The oceans of the world carry so much gold in suspension that if mining it to its fullest were possible, enough would be harvested to give each person on earth about nine pounds of solid gold. Scientists know each cubic mile of seawater holds about 25 tons of gold. That's a total of 27 million tons of gold in the world's oceans. Although this is an impressive amount, so is a cubic mile of ocean. The gold is dispersed throughout the water, present in such minute concentrations that no method has yet been devised that can extract it profitably.

Imagine, for example, a modest-sized room 10 feet wide, 12.5 feet long, and 8 feet high. These dimensions represent a volume of 1,000 cubic feet. Filled to capacity with seawater, the room would contain only about five one-thousandths (.005) of an ounce of gold. That is certainly not enough to get excited about, and of course not enough to try to extract. Only about five ounces of gold can be found in a billion ounces of seawater (one part per 200 million). So don't bother packing your gold pan for the next ocean voyage. All of the many scientific attempts to recover the minute flecks of gold strewn throughout the sea have failed.

At present the only metal successfully extracted from the sea in commercial quantities is manganese. It occurs as nodules carpeting the ocean floor in an almost continuous layer, so in many places it is easy and economical to recover. The manganese appears to be precipitated from the water and forms over other mineral deposits; within the nodules are about 40 different minerals and an occasional shark's tooth, whalebone, or piece of pumice. The nodules are valued more for the other minerals (cobalt, zinc, nickel, copper) than for the manganese.

About 85 percent of all substances dissolved in the oceans is sodium chloride, the mineral halite, known best as common table salt. Each cubic mile of seawater contains over 166 tons of dissolved salt, enough to supply the world's demand for several years. Scientists have calculated that if all the salt were extracted and spread over the land it would form a layer more than 500 feet thick!

In some places salt is still recovered from seawater by simple evaporation, the same method used by the Chinese as early as 1000 B.C. Most salt, however, is mined from brine wells and salt domes, which show that the continental area where they are found was once covered by oceans. Being cut off from the sea the water was subject to evaporation, and the mineral matter carried in solution eventually precipitated out, yielding in many cases enormous concentrations of salt ready for the dinner table. Of course this is an oversimplification of the complex mining methods that are applied, but salt is salt.

Phosphate, the commercial product from phosphorite, is used extensively as a fertilizer and, because it is much in demand, is a valuable economic commodity. Scientists know that a rich deposit of phosphorite lies on the seafloor off the coast of San Diego and that economic recovery is quite feasible. However marketable the deposit may appear, it will never be mined because the area was once a naval firing range. Thousands of live shells and mines are spread out over the phosphorite waiting to explode when even slightly disturbed. The deposit has been thoroughly studied by scientists of many specialties, and all have concluded the risk is so great that this will remain, possibly forever, a lost treasure in the sea. Several people were heard to comment, as they walked away from all these riches, that such are "the fortunes of war."

The Windiest Place in the World

As every schoolchild knows, Antarctica is the coldest continent in the world. However, this frozen landmass boasts another claim to fame.

Commonwealth Bay, at George V Coast just south of Australia, has been named by scientists as the windiest place in the world, the "Home of the Blizzard." Here masses of dense cold air from high on the Antarctic Plateau slide downhill toward the coast. The shape of the land tends to funnel and therefore concentrate the winds so that they often accelerate to speeds of over 200 miles per hour. These are hurricane speeds. And make no bones about it, this wind is cold. In fact the chill factor defies the imagination, making tourist trips to Adelie Land quite infrequent.

Floating Water

About three billion years ago, after more than one and one-half billion years of cooling down, the earth became blanketed with a dense cloud that included much hydrogen and oxygen. The cooling persisted until the gases that surrounded it condensed into water and fell as rain. The downpour continued unabated for about 60,000 years, filling ocean basins and lowlands with an estimated 326 million cubic miles of water. And no matter how much is used, polluted, or wasted, the earth retains all of the water ever created.

At present over 70 percent of the earth's surface is covered by water, 97 percent of which is contained in the oceans. Of the remaining 3 percent that is freshwater, over 2 percent is held in ice caps and glaciers in high mountains, and less than 1 percent is in reserves below the earth's surface, in the atmosphere, and in rivers, lakes, and inland seas.

Most of the world's great river systems, from the Amazon to the Ganges, the Rhone to the Columbia, originated in glaciers. Nearly three-fourths of all the freshwater in the world, just under seven million cubic miles, is stored in the form of glacial ice. Authorities estimate that this reserve is the equivalent of about 60 years of rainfall over the entire globe.

Greenland and Antarctica alone are capped by five million cubic miles of ice. If global warming proceeds as predicted, this ice will gradually melt, causing the sea level to rise nearly 300 feet worldwide and adding to the percentage of saltwater. And coastal cities such as Boston and New York will be visited only by persons in scuba gear or on a submarine.

During the ice age that ended 10,000 years ago, over 30 percent of the land surface was covered by ice sheets thousands of feet thick. This moving glacial ice constantly supplied the coastal seas with gigantic icebergs. Then, as now, when the ice moved from the land into the sea, these enormous chunks of ice would break off, a process called calving. They float away from the shore, and thus are born icebergs. Considering that so much ice covered the lands during the last ice age, the northern seas must have been heavily dotted with massive icebergs. These ice age icebergs may have reached as far south as the latitude of Mexico City, well within the Torrid Zone. This would indicate that the oceans maintained a rather frigid temperature.

Since ice is so heavy and solid, observers of icebergs, and of ice cubes hugging the surface of cool drinks, wonder just why ice floats on the very water from which it came. The behavior of water as it cools toward the freezing point is amazing. Almost all materials contract as they cool, reaching their greatest density at freezing. Not so with water! It contracts as it cools down, but only to 4 degrees centigrade; at this point it begins to expand until it freezes at 0 degrees. Because expansion decreases its density, the ice that results takes more space than the water from which it came.

A particular volume of ice is lighter than an equal volume of water. A cubic foot of water weighs 62.5 pounds; a cubic foot of ice weighs

56.9 pounds, a difference of about 5.5 pounds. As do all lighter substances, the lighter ice will float on the heavier water. Moreover, since the ice consists of freshwater (which is lighter than seawater) the icebergs float partly above the surface.

The physical change of water into ice creates a force so powerful that water pipes break, car radiators crack, carbonated beverages explode, and caps on bottles of frozen milk sit on an ice tower several inches high. Great chunks of rock come tumbling down mountainsides, split off by the expansion of freezing water between the cracks in the bedded rock (a process known as frost wedging). The result of this characteristic of water is that lakes and rivers freeze at the surface instead of from the bottom up. The living creatures remain comfortable in the relatively warm waters, even when the surface of the lake or river is frozen solid. This fact accounts for all aquatic survival, perhaps all life!

Another important property of water is its ability to absorb free oxygen; this increases as the water temperature decreases. One would assume that warm tropical waters are the home of most sea animals, just as tropical forests are more densely populated than polar regions. Actually cold waters contain the most abundant sea life because cold water tends to hold more oxygen than warm. In fact the solubility of oxygen in water at 0 degrees centigrade is approximately twice that of water at 30 degrees; for this reason, marine life becomes more abundant as seawater gets colder. Surveys show that sea animal populations increase greatly toward the poles and lessen near the equator. Exceptions occur where upwellings of cold water exist.

The photographs one often sees of a warmly dressed angler fishing through a hole bored in the ice, with a large catch of fish sitting alongside, are usually not fabricated. The fish beneath the ice are hungry and very energetic. Invigorated by waters abundant in dissolved oxygen, they are always ready for the next meal.

Birth of a Legend

Almost all legends, in contrast with myths, are accepted as having some basis in fact. Just how much is fact is often difficult to discern because, with time and many retellings, the original facts can be completely inundated with distorted details. An excellent example is provided by the familiar Greek legend of Jason and the Golden Fleece.

Jason was the son of a king of Thessaly. With his band of 50 heroes, the Argonauts, he set out to find and capture the Golden Fleece. This hide of a golden sheep hung in a sacred grove in the kingdom of Colchis, located on the eastern shore of the Pontus Euxinus, known today as the Black Sea. Here the fleece was guarded by a sleepless dragon.

A dashing warrior, Jason managed to enlist the aid of Medea, a sorceress and princess of Colchis. Jason performed incredible feats that included slaying the dragon who guarded the gilded treasure. He seized the Golden Fleece and, after further adventures, escaped with it back to Thessaly accompanied, of course, by Princess Medea. A typical legend, recounting the audacious exploits of its clever, invincible hero, the story of the Golden Fleece has a definite basis in fact. The original story occurred in ancient Colchis.

Scholars have long been aware that in Colchis there was a tribe known as the Tibareni who practiced a form of placer mining by sluicing rich, gold-bearing stream gravels over unscoured sheepskins. The skins and their natural oils caught and held on to the particles of gold. After shaking out the coarser nuggets, the ancient gold miners would hang the fleeces on trees to dry. When the skins were thoroughly dry the fine gold dust was beaten out of them. The citizens of Colchis accumulated a vast amount of gold, and thus was born the Golden Fleece. Many Jason types of expeditions were launched to find a golden treasure; few did.

So who was Jason? Throughout history there have been many Jasons. He was the prototype of the California forty-niners, the sourdoughs of the Klondike, and all the gold seekers through the ages. Under whatever name, the searcher was for the moment filled with the spirit of Jason and was in truth searching for the Golden Fleece.

Recent evidence confirms that most of another legend is true. In early 1990 a team of scientists scouting the boundary between two ancient Greek cities unexpectedly uncovered an inscribed monument previously known only through the writings of the first-century Greek historian Plutarch.

Plutarch had written an account of the legend. But he was known to invent dialogue in many of his writings, so with the lack of any corroborating evidence his account was not taken seriously. In Plutarch's "Life of Sulla" he describes how two townsmen of Chaironeia made possible a great victory for the Roman general Sulla.

Troops from Pontus, a kingdom near the Black Sea, had camped on a floodplain north of the city of Chaironeia, and one detachment perched on a hill known as Thourion. According to Plutarch's tale, Sulla had positioned his foot soldiers between the main body of the forces from Pontus and the town of Chaironeia but was absolutely unable to repel the hilltop squads in frontal attacks. His losses were heavy, and his campaign appeared hopeless. Finally two townsmen named Homoloichos and Anaxidamos offered to lead Roman soldiers up a back pathway on Thourion, thereby surprising the invading Pontus soldiers. Plutarch wrote that the plan worked perfectly; at least 3,000 infantry from Pontus were destroyed in a surprise attack on Thourion's rocky slopes. This allowed a successful Roman attack on the river troops, resulting in a complete victory for Sulla.

In honor of the heroic townsmen, according to Plutarch, the grateful general erected two stone trophies to celebrate the victory, one on the plain and the other on Thourion.

Classical scholars knew that such a battle may have occurred on a hill called Thourion. On the basis of Plutarch's description of the event, they selected several hills about ancient Chaironeia as possible battle sites. However, much doubt was placed on his accuracy and, for that matter, on all of his talents except the spinning of tall tales.

In February 1990 an archaeologist and four graduate students from the University of California, Berkeley, found the long-lost legendary monument atop a hill near the archaeological site of Chaironeia which was, incidentally, Plutarch's hometown. Near the base of the hill they uncovered about 150 stone blocks that appear to be the remains of a temple dedicated to Apollo and built during Plutarch's time. But most important was the discovery of the ruins of a shrine on top of the hill. This hill could now be identified as Thourion, for amid the pile of rubble was a marble block, about three feet wide and one foot high, inscribed with three words "*HOMOLOICHOS,*" "*ANAXIDAMOS,*" and "*aristis*" (the Greek word for heroes).

On at least one occasion, Plutarch was not inventing dialogue. Almost the entire legend is now accepted as historical fact.

A Nose Is a Nose

The world as perceived by a dog is quite different from the one with which humans are familiar. The dog's eyes can't distinguish hues, only tones, so they see a world of black, white, and various shades of gray. The human eye has two kinds of light receptors: rods and cones. The cones are color sensitive, so humans can recognize 120 to 150 different hues. The rods take over in the dark and do not distinguish color or details. This is the world the dog sees.

Seeing means far less in a dog's life than it does in a human's, because other senses take over. A dog's hearing is about 140 times more acute than the auditory sense in humans, and scent discrimination is thousands of times more keen. It can be said that animals with scent discrimination smell their world; humans see theirs. Humans can no more imagine the vivid range of sensations revealed to a dog by its highly developed olfactory sense than a polar bear might imagine the landscape of an Amazon jungle.

During the early 1960s a series of famous experiments was conducted at a leading university to determine the sense of smell possessed by dogs. The scientists used iodoform (a compound of iodine analogous to chloroform), selected because of its distinctive odor. They found that, although the sense of smell varied among certain breeds, all dogs showed an extraordinary ability to detect odors. The average dog could detect iodoform in a solution of one part to four million, even when it had been disguised with four or five other powerful scents.

Packed with many times more olfactory nerves than are allotted to the human nose, the dog can pick up the distinguishable odor of every living thing, animal or vegetable. Since each human being smells different to the dog, it will not mistake stealthy bungling burglars for its faithful human companions. The dog can also detect variations in odor resulting from emotional or physical conditions such as illness, joy, or fear. When it smells fear, the dog may be expected to take advantage of the adversary. Freshly fallen snow and dampness hold scent very well, so dogs are efficient at sniffing out victims of avalanches, floods, or tornadoes.

Dogs and humans share one feature, although they display it differently. Each human may be identified by his or her fingerprints, since no two people, even identical twins, have the same pattern. Each dog also has an identifying characteristic, but again the nose reigns supreme, for each noseprint is unique. And if we recall how the dog recognizes each person it meets, we may realize that the odor of a human may be as identifiable as a fingerprint.

CHAPTER FIVE

Paleopathology

For years many museum curators have kept special collections of fossil bones deformed by some disease or injury. Only recently have scientists actively begun to study such collections, creating a new field of research called paleopathology. These studies add much to existing data on how ancient animals looked and moved. Deformed remains offer clues concerning their lifestyle, how they interacted, and who roughed up whom.

A leading museum in Riverside, California, has a fossil leg bone of a deer that died seven million years ago. The leg shows a definite healed break. Apparently, while running at top speed the animal stepped into something like a gopher hole and broke its leg. The momentum of running would have kept it in motion, and the broken upper part of the leg bone slid down the side of the lower part of the leg. The broken ends are about two inches apart, and in this position the two bones fused together. The animal must have spent the remainder of its life with one leg significantly shorter than the other three.

Since carnivores always seek out the weak and disabled, this severely disabled animal's survival until its broken limb healed seems almost miraculous. After the healing it would have walked with a very noticeable limp. This otherwise healthy young deer was unable to survive very long with a permanent disability. Paradoxically, the very bone that disabled the animal was the only one that survived its predator's feast.

At the Tyrrell Museum of Paleontology in Drumheller, Alberta, the fossil collection contains a large number of damaged bones belonging to a particular family of dinosaurs, the hadrosaurs. Probably more is known about how this group of dinosaurs lived than any other because of the frequency of injuries exhibited by their bones. Commonly described as duck-billed herbivores, they lived during the last part of the era of dinosaurs and died out about 66 million years ago. An examination of their skeletons displayed frequently broken ribs that later healed, indicating that they had survived the encounter. These injuries appeared only in larger, probably adult male, skeletons. Similar rib injuries show up on hadrosaur skeletons at the Royal Ontario Museum in Toronto.

Scientists believe these injuries were the results of severe kicking between males as they fought for the attention of females during mating season. A violent kick from a four-ton dinosaur could be quite damaging; one skeleton showed evidence of a kick that broke six ribs at once. Along with mating rituals, fights between males may have been the means of establishing dominance in a herd. Males seem to have kicked at each other with their oversize hind feet in the manner of modern kangaroos.

Another injury common to hadrosaurs and many other dinosaur groups, including the terrible *Tyrannosaurus rex*, is broken vertebrae at the base of the tail where it attaches to the hindquarters. This injury is found mainly in dinosaurs that appear to have been females. The tops of the vertebrae are compressed and cracked as if a great weight had pressed down on them. The vertebrae often healed in their broken position.

At the risk of offending modest colleagues, a paleontologist demonstrated with illustrations how a larger male pressing down on a female during mating could crack these particular vertebrae. Since the vertebral cracks were not fatal, they were not counterproductive to the mating process. Mating was nevertheless a painful affair for the female dinosaur. Although she survived the encounter, she probably dreaded the next mating season.

Another injury common to herd animals such as the hadrosaurs was numerous cracks toward the end of the tail vertebrae. Here the bones were bent and broken as if they had also been squashed by a great weight. Scientists know that these animals, probably not the most graceful creatures, traveled in great herds numbering in the thousands.

One hadrosaur in a resting position with its tail on the ground would be at the mercy of whatever awkward giant came lumbering by. Judging from the many such injuries in the fossil record, having one's tail stepped on must have been an ever-present hazard. Even though hadrosaurs have traditionally been portrayed as peaceful, mild-mannered animals, the injuries suggest that they were careless, aggressive, or accident prone—possibly all three.

Researchers have also detected stress fractures in the toe bones of several groups of Ceratopsian dinosaurs, the family of horned dinosaurs that includes the mighty triceratops. This living, moving tank weighed 8 to 10 tons. Stress fractures are not common in four-legged animals and usually occur when extraordinary demands are placed on the bones for brief, intense periods. In the case of the triceratops, the toe fractures imply that the great beasts were in the habit of stamping their feet. Scientists believe that this gesture was part of a challenge as one male defied another for the charm of the female or that it signaled a confrontation to determine leadership of the herd.

As the two living tanks squared off at each other and stamped out their challenge 80 million years ago, the ground must have shook under the impact. Then with heads lowered, the two tanks charged each other. The repercussions of 20 tons of bone and muscle colliding would have reverberated through the primeval surroundings.

More and more physicians study ancient injuries and diseases to gain valuable insight into modern physical disabilities. One specialist in arthritic diseases teamed up with a paleontologist at the Kansas University Museum of Natural History. They examined the bones of a mosasaur, a large sea lizard that lived about 75 million years ago. One of the tailbones appeared to the doctor to be diseased; it was gnarled in appearance, with tiny bumps and holes. The physician, who had seen many bones like this, presumed correctly that the holes eaten into the bones were the result of viral or bacterial invasion.

Slicing the suspect vertebra into sections, the scientists found, in addition to healthy bone, bone with blotches, swirls, and squiggles; it looked more like bone soup than bone structure. The sea giant had definitely been infected. The cause of the infection, found deeply imbedded in the bone, was a fingernail-sized tip of a shark tooth. As the scientist remarked, "A dirty-mouthed shark had taken a bite out of this mosasaur." From the tooth they were able to identify the shark as one of several species that lived in this inland sea. This body of water, a thousand miles wide and stretching from the Gulf of Mexico to the Arctic Sea, covered the area where the state of Kansas exists today.

The two scientists also examined what appeared to be perfectly

healthy tail vertebrae. Within the bone marrow, bands of decay wove through the delicate lacework pattern of the normal bone. The physician immediately recognized the first evidence in any animal other than humans of a vascular necrosis, tissue death due to a lack of blood supply. In simple language, the mosasaur had the bends!

Bends occur in human divers who rise too rapidly. The increase in pressure that accompanies a diver's descent causes nitrogen gas to dissolve in the blood. As the diver returns to the surface the gas precipitates out again. This is harmless if the diver's ascent is gradual, but in a rapid ascent the pressure changes too quickly, and bubbles of nitrogen form in the blood, blocking small blood vessels and cutting off circulation. This is extremely painful and sometimes fatal.

Scientists have long accepted that this species of mosasaur dived to great depths in the ocean, but a diving lizard should have known how fast to surface to avoid excruciating pain. In this case, the ascent was extremely rapid perhaps because, as the evidence shows, a hungry shark was on the mosasaur's tail. Continued research also uncovered a giant prehistoric sea turtle with a similar problem. Teeth marks on the bone show it had been pursued by a mosasaur that managed to grab one of the turtle's legs and bite it off. The turtle, whose wound healed completely during its lifetime, must also have ascended rapidly, with a hungry mosasaur in pursuit.

The researchers went on to confirm that mosasaurs throughout the seas of the world were subjected to the bends. Despite the excruciating pain they must have endured during a hasty retreat, suffering the bends did not prove fatal. And their rapid though painful ascent helped them to avoid a fatal encounter with a hungry shark. The mosasaur, as well as the sea turtle, never did evolve the decompression resistance that has protected other types of marine reptiles. A relative newcomer to the ancient seas, the mosasaur existed for only about 25 million years, becoming extinct along with the dinosaurs at the end of the Cretaceous Period.

At a recent medical meeting at the Field Museum of Natural History in Chicago, the physician delivered a lecture on paleopathology. Afterward a member of the audience brought him a piece of the backbone of an 11,500-year-old bear. It had been discovered in Indiana and was streaked with strange scars. The physician took the bone to his laboratory, where x-rays revealed familiar-looking bone damage. Medical tests and immunoassays performed on the bear bone confirmed his diagnosis, and he moved paleopathology to a new plateau of precision and validity; even in fossilized bone the original bacteria, antibodies, and soft tissue were responsive to the immunologic tests.

Fortified by unequivocal results of these tests and immunoassays, the paleophysician presented his bear findings to a community of his medical peers. With courage and conviction he was able to announce, "This bear had syphilis!"

Never Insult a Priest

A Persian physician by the name of Rhazes (c. 854–925), drawing on his own experience, wrote the first accurate description of measles and smallpox. He experimented with animal gut and found it worked well for suturing, and he recommended the use of plaster of Paris for casts. He was also the first physician to realize that fever was a natural defense mechanism, the body's way of fighting disease.

Rhazes learned, practiced, taught, and wrote about medicine and surgery with great authority. Having journeyed through much of Africa and the Arabian peninsula, he became an unchallenged teacher of the healing arts merely by saying, "According to my own experience . . ." He wrote over 200 treatises on medicine; his short, concise works and those with well-arranged topics were used for centuries as medical textbooks. One of his works, *Continens Liber* ("The Continent of Medicine"), was a compendium of everything known about the art of medicine. Without method or arrangement, it was a vast undigested mass of information in huge folios. Only two complete copies of the bulky manuscript were ever reproduced. No doubt the cost was as prohibitive for any buyer as the amount of time a calligrapher would have spent copying it.

Continens Liber made Rhazes immediately famous, but it terminated his practice of medicine. The book thoroughly offended a mullah, a learned Muslim priest who was able to interpret theology without fear of contradiction. Apparently somewhere in his massive book Rhazes managed to contradict the mullah. The infuriated wise man ordered the doctor beaten over the head with the manuscript until one of them broke. Unfortunately Rhazes' head broke while the ponderous manuscript remained intact. The result was permanent blindness for Rhazes and the end of his medical career.

Lion Trouble

On the evening of August 23, 1972, a healthy but hungry lion pushed his way through a thornbush fence surrounding a Masai village while the rest of the pride waited outside. Three eight-year-old boys were sleeping in the open next to their goats. The instantly alert goats fled, but the children slept soundly, much too soundly. The hungry lion seized and quickly killed one of the boys; this awakened the other two who scrambled to safety while the victim was being consumed on the spot. The boy's family retaliated by killing a zebra and lacing its carcass generously with poison. Twelve lions feasted on the meat and died; the man-eater was assumed to be among the unlucky dozen.

Since their arrival in the Kenya-Tanzania area in the 17th century, the Masai have been acclaimed as a fierce warrior people whose acts of bravery were unsurpassed. One of their most daring pastimes, for which they trained as a sport and followed elaborate rituals, was hunting the lion with only a spear and a dagger. Use of any other weapon, such as a firearm, was considered cowardly and unfit.

Learning to deal with lions was essential for the Masai, since they subsisted basically on the herds of cattle that they guarded vigilantly. Their herds were considered gifts from the gods, and individual wealth was, and still is, measured by the number of cattle a person owned. Cattle provided their principal foods—fresh blood, milk, and meat— and their building materials—dung and hides. Whenever a marauding lion attempted to dine on one of their cattle, the entire village of warriors would take up knives and spears and engage in a death-dealing lion hunt. The lion was hunted down and encircled by platoons of warriors and, as the circle closed, it would attempt to fight its way through.

Although the confrontation usually ended with the lion's death, one or more of the warriors often went with it. The hunts were quite hazardous, especially when more than one lion was caught in the ever-shrinking circle.

Because the government finally outlawed such deadly encounters, lion spearings are no longer part of the Masai way of life. But domestic cattle are still easy prey for lions, whose acquired taste for beef has been sustained by villagers' removal of dead and diseased cattle to areas beyond their compound. The lion typically tries to avoid a confrontation. In a well-planned maneuver, it will enter the village, grab the prey, and carry it off. This is not an easy task for a 400-pound lion with its teeth around the throat of a cow weighing twice that. Nevertheless, by deftly sliding under the carcass, the lion shifts the weight onto its back and holds its tail rigid as a balance. In this way an escaping lion can jump over an 8- to 12-foot thornbush barricade surrounding a village. The Masai agree that without the tail for balance, such a feat would be impossible.

Lions generally tend to avoid contact with humans; usually the old, handicapped, or very hungry lions seek humans for food. However, between 1932 and 1947, lions killed an average of over 100 people a year in just one area of Tanzania. During that time not a single lion was killed in retaliation, or even in self-defense, because Tanzanians believed that man-eaters were being directed by the spirit of a deceased person who returned to settle old grievances. Therefore, to kill one of the great cats would cause countless deaths among the people.

When a game warden tracked down and shot a lioness that had just killed a villager the tribesmen believed many of them would be doomed, so they shunned the man. But as time passed and no harm came to them they started to wonder. Finally several of the bolder villagers began to defend themselves and even to hunt the lions, and the ranks of the lion thinned under the spears and firearms of the vengeful natives. Human hunters also paid a price for their attacks, particularly when, armed with old, unreliable, low-caliber shotguns, they only wounded the lion. Even with vital organs shot away, lions have been known to continue charging until they topple their attackers, leaving them either badly mauled or dead.

Several years ago a park official driving on a road between Nairobi and Voi National Park passed two natives he recognized as relatives of a local tribal chief. He turned around after about three minutes when he recalled that he wanted to give them a message for their chief. He found no trace of the men he had passed just a few minutes before but, assuming they had been given a lift by some vehicle, forgot the incident. Later that afternoon he received a report that two badly chewed bodies had been found in the district. The unfortunate men had been victims of a pair of man-eating lions.

The drag marks and tracks told the story. The lions had attacked the men while they were still on the road, dragged them just 30 yards off the roadside, and proceeded with their meal. All this took place during the eight minutes between the officer's first sighting of the men and his return to give them a message. The lions were probably crouched in the bushes having dinner while the park ranger made several runs up and down the road in search of the Africans.

Park officials discovered that a large pride of lions, deciding humans were the best food for their delicate stomachs, had proceeded to kill and devour local people with apparent relish. The government hired many professional hunters who, although they did take their toll on the lions, were no match for cautious and clever cats. The villagers finally abandoned their homes and moved into Nairobi or other less threatening sections of the country. The lions also moved on; their predatory activities continue but in more remote areas.

The Naming of the Horse Latitudes

The ocean regions from about 30 to 35 degrees latitude in the Northern and Southern Hemispheres are known as the horse latitudes. Off the North American coast, these would be the latitudes from northern Florida to Cape Hatteras, North Carolina. Here the prevailing westerly winds move toward the poles and the trade winds (easterlies) move toward the equator, leaving an area in these latitudes where the weather is warm, clear, and calm. Old sailing ships often spent a lot of time just sitting on a smooth, glassy ocean waiting out the all-too-pleasant weather.

On 18th-century trade ships, remaining motionless on a becalmed ocean had anything but a calming effect on the sailors. During that century horses were frequently transported from New England to the West Indies. In the 30- to 35-degree north latitudes, a speedy, profitable voyage often became a worrisome wait while provisions dwindled, perishables perished, and supplies of water and fodder diminished.

Rarely was a ship prepared for a delay of more than a few days. If it was becalmed for as long as a week, some sacrifices had to be made.

What would be sacrificed? The horses, naturally paying the ultimate price for the lack of wind and the exhausted supplies of livestock food, they were thrown overboard and quickly became food for the sharks.

In honor of the sacrificed horses, sailors and shipowners named the area of calm sea the "horse latitudes."

Oology

The number of "ologies" seems to be endless. Over 300, from *actinology* through *zymology*, are listed in the *Complete Rhyming Dictionary*. A few of the more common "ologies" are geology, paleontology, anthropology, ornithology, and zoology, which mean, respectively, the study of the earth, fossils, humankind, birds, and animal life.

Oology refers to the study of eggs. Considering how fragile and inaccessible they are, eggs are not most nature lovers' first choice for collection and study. Nevertheless the study of eggs and nests has claimed, among its ardent enthusiasts, some of the United States' eminent ornithologists and conservationists, including Presidents Theodore Roosevelt and Franklin Roosevelt.

Collecting eggs has its hazards and mishaps. In 1889, collector Maurice Thomson discovered a rare clutch of ivory-billed woodpecker eggs. They were in a nest on the face of a cliff, about 30 feet above the ground (considered easily accessible to oologists). Thomson climbed to the level of the nest, where he stood resting and admiring his find. Just as he reached for the eggs, he felt an itch on his hip. Twisting around to scratch, he accidentally knocked the eggs out of the nest and watched in anguish as they all fell to the ground and smashed. He later remarked that all he could think of at the time was, "Humpty Dumpty had a great fall. . . . All the king's horses and all the king's men couldn't put Humpty Dumpty together again."

One of oology's most memorable stories is that of Major Charles Bendire, egg collector and Indian fighter. In 1872, while on patrol in central Arizona, he noticed through binoculars a zone-tailed hawk's nest high in a tree. Leaving his troops to set up camp, he rode to the tree, tethered his horse, and climbed to the nest, keeping a wary eye open for Indians and concealing himself as much as possible.

From the nest, he plucked one of the eggs. Caution escaped his mind as he marveled at this incredible addition to his growing egg collection. An Apache scout quickly spotted him and got off a snap shot with a carbine. As the bullet zipped harmlessly over the major's head, he reacted instantaneously. Shoving the egg into his mouth for safekeeping, he hurried down the tree, jumped onto his horse, and galloped wildly back to camp with several Apaches in fervent pursuit. He managed to reach the camp, where a brief, pitched battle drove off the Apaches.

Then the real problem began. As he rode headlong into camp, gasping and gagging, Bendire discovered that he couldn't spit the egg out. It seems that as he had tried to avoid biting the egg, his jaws had tensed up and swelled. He simply could not open his mouth wide enough to remove the egg. Several men, under threat of court-martial, pried open his jaws and got the egg out intact. Although they did break one of his teeth, Bendire thought it a small price to pay for a perfect, uncracked egg of a zone-tailed hawk.

Charles Bendire later became the first curator of oology at the Smithsonian Institution, where the storied egg survives to this day, along with about 130,000 others.

Another prodigious collector was Wilson C. Hanna, who collected eggs from every continent. To remove some of the hazards of getting to nests and eggs, Hanna fabricated a folding metal ladder that could be attached to a tree or a cliff in pursuit of an elusive prize specimen. His collection of over 30,000 sets of eggs was donated to the San Bernardino County Museum, although a few rare specimens have been shared with the National Museum of Natural History in Washington, D.C.

One of the rarest of eggs was added to Hanna's collection at the museum by Dr. Warren D. Mateer of Redlands, California. It is an egg of the extinct elephant bird found on Madagascar. The largest and sturdiest of any egg of any species yet discovered, it holds almost two gallons of liquid. Or, in the terminology of the oologist, it would take the contents of 24,000 eggs of the vervain hummingbird of Jamaica to fill the egg of the elephant bird!

Robinson Crusoe, a Scottish Pirate

Daniel Defoe's all-time favorite, the classic story of Robinson Crusoe, might never have been written but for an argument. The argument took place at sea between a Scottish privateer and his captain one night in 1704. The privateer, Alexander Selkirk, was the ship's first mate on an expedition to the South Seas. No one seems to know what the argument was about, but in the heat of the exchange Selkirk demanded to be put ashore. He was—on one of the uninhabited Juan Fernandez Islands, off the coast of Chile. In all, Alexander Selkirk could have considered himself a lucky man to have gotten off with just abandonment; in those days the captain at sea was an absolute monarch and could have had the ship's first mate flogged or even hanged for possible mutinous actions.

So Robinson Crusoe was based on a real man, Alexander Selkirk, who lived the lifestyle described by Daniel Defoe as a man shut off from civilization for many years. Selkirk raised goats and other small animals and had ship's dogs as his everyday companions. He did not encounter cannibals such as his fictional counterpart did, nor did he have his man Friday. Perhaps if he had stayed on the island longer than four and one-half years he might have. Rescued in 1709 and taken back to England, Selkirk recounted his adventures. This aroused much public interest, particularly among writers looking for a good story.

Daniel Defoe was a journalist at the time. He sought out the rescued pirate and in successive interviews came up with the immortal story of Robinson Crusoe, a pious man who was marooned on an uncharted island for over 24 years.

The Rounding of the Earth

In the spectacular repair mission of the defective Hubble telescope, fascinating space walks were transmitted for the world to watch. The views were amazingly clear, and in the background are occasional glimpses of earth, very round and spherical looking, like the big blue marble described by astronauts.

Early viewers of the heavens could be excused for their myopic interpretation of the earth and sky. Their understanding was based on naked-eye observations and was limited by small-range wanderings on the home planet. Thinkers in most early civilizations accepted the earth as a flat, unmoving center of the universe, with sun, moon, stars, and planets circling around. Among most ancient peoples the view of the universe was embellished by religion: Egyptians regarded the earth as a flat square under a pyramid-shaped sky; Greeks placed their country in the center of a flat, circular earth with heavenly bodies circling around it; the early Hindu concept of the world was a plate resting on the backs of four massive elephants standing on a giant motionless turtle.

Many ancient scholars who observed and studied the heavens were able to delineate stars from planets, trace their motions and cycles, and draw some conclusions. No one knows who first imagined that the earth was round, but the idea didn't wait for Columbus. It may have been at the school of Pythagoras (580–500 B.C.) in a Greek colony in Italy that the theory of a round earth became a firm belief. This elite company of Greeks could easily observe how ships vanished over the horizon. Travelers related to them that as they sailed north new stars rose in the sky and those stars in the south dropped from sight. To members of the colony this was indisputable evidence that the earth was round. The Pythagoreans developed the idea of a round earth into a complete panorama of the universe, with the central earth surrounded by transparent spheres that carried the sun, moon, planets, and stars, all whirling in perfect harmony.

By the time of Aristotle (384–322 B.C.) many of the Greek thinkers were convinced that the earth was round. Aristotle himself pointed out that the round shadow of the earth on the moon during eclipses clearly showed the spherical shape of the earth. Even in the face of their advanced reasoning, Aristotle and his followers accepted the earth as the center of the universe, with the sun and all the planets revolving around it. Several questions remained unresolved. There was no explanation of why the planets moved slowly from west to east, while stars would have to fly at incredible speeds from east to west to make the trip once per day.

These mysteries were answered by the Greek astronomer Aristarchus of Samos (325–250 B.C.). He believed that the stars did not spin every day but that the earth did the spinning, and this caused the stars to appear to move across the sky at night. He also believed, correctly, that the earth traveled once a year in a huge orbit around the sun, as this

would explain the change in star patterns throughout the year. He reasoned that the planets seem to move backward because they circle more slowly than the earth. Although this explanation was somewhat flawed, his reasoning that the planets (including earth) revolve at different speeds was accurate. And he placed the sun in its correct position at the center of the solar system.

Few people of the day agreed with Aristarchus. They labeled him a hoax, fraud, or lunatic, most likely because they resisted giving up the earth's position in the center of the universe. His fellow astronomers argued that if the earth orbited the sun, one would be able to observe parallax shifts in the stars.

Parallax shifts, which seem incredibly sophisticated for astronomers whose observations of the heavens were seriously limited, are easy to test. Holding a finger at arm's length in front of a contrasting background, one can see a motionless finger "jump" back and forth by looking at it first with one eye and then with the other. The finger appears to move to different parts of the background because of the distance, however small, between the two eyes.

Aristarchus argued that the stars were too far away from the earth for any parallax shift to be visible. Again he was correct, because slight parallax shifts are detectable only with modern telescopes.

With all their logic Aristarchus's theories fell on deaf ears, and the Greeks continued to place the earth at the center of the universe. His hypotheses were ignored by Ptolemy (A.D. 90–170), whose concepts of a geocentric (earth-centered) universe endured for 1,400 years. The traditional view of the solar system remained unchallenged until the Polish priest Nicolaus Copernicus (1473–1543) developed some radical notions about the order of heavenly bodies. Though he admitted that the idea was absurd, Copernicus began to think of a motion of the earth. As humbly as possible, so as not to offend the religious hierarchy of the day, he suggested that "the Sun, as if on a royal throne, rules the family of planets (including the godlike Earth) as they circle around him."

Copernicus's original manuscript credited Aristarchus with the shocking concept of a solar system 1,800 years before. Fearing the outrage of church leaders, he withheld publication of his book until he was over 70 years old and near death. His book, *De Revolutionibus*, was banned by the Church for over 300 years, but he did not live to suffer through the centuries-long battle.

Though the earth has been acknowledged as a spherical planet for many centuries, and views from space show its rounded outline unequivocally, we still have the Flat Earth Society.

So Far, yet So Near

In the year 1970 a geologist with a field crew was conducting a geologic survey in the mountains just outside of Cheyenne, Wyoming. A desert geologist, he was not at home in the mountains, where precipitation was relatively high. It was during the winter, and weather forecasts warned of a heavy snowstorm.

The men were deeply involved in their work when it began to snow. Within a short time a raging blizzard was upon them, and none of them could see more than a few feet in front of themselves. They piled into their jeep, drove it almost immediately into a snowdrift, and had to abandon it. Their sense of direction completely gone, they walked holding on to each other's shoulders to keep together. In a short time the lead man stumbled onto a barbed-wire fence. Knowing that fences go somewhere, the men held on and went in a direction that seemed downhill. It led to the highway, which was probably only a few hundred yards away. Before the men had time to wonder where to go next, a truck picked them up and drove them to Cheyenne. They were lucky.

Scientists know that prolonged exposure to a blizzard will cause giddiness and a lost sense of direction, both of which the geologic crew experienced. A similar incident occurred in Switzerland during the same season that the geologists in Wyoming experienced their winter adventure. A man driving home from work was suddenly caught up in a blinding blizzard. Almost immediately he drove his jeep into a snowdrift and had to abandon it. Within a few steps the man was hopelessly lost. Luck was with him as it had been with his counterparts in Wyoming, because he found shelter by just stumbling onto it. The home he approached belonged to a neighbor, and he was immediately taken inside to the warm fire. When the storm abated he was amazed to find his jeep in a drift just a few yards from the front of his own house.

Forever Beautiful

In the name of vanity, Americans, both men and women, spend over $5 billion a year on cosmetics, beauty salons, and barbershops. Artificial adornments of the face and body have been a part of the human experience for many thousands of years; painting, powdering, and perfuming have been in existence since early Stone Age cultures.

Researchers have strong evidence that our Paleolithic ancestors used cosmetics: the abundance of red ocher at many early sites suggests that coloring of faces and bodies was an established ritual. Although the short, stocky Neanderthal female may have resembled a wrestler too much to appeal to modern tastes, her painted body may have appeared all the more attractive to the sturdy, brutish males of 100,000 years ago. Facial and body decoration may have begun as beauty marks for the Cro-Magnon of 25,000 to 30,000 years ago, but they became a serious feature in rites of religion and warfare. Such practices continued into the age of civilization; archaeologists have unearthed palettes for grinding and mixing face powder and eye paint dating as far back as 6000 B.C.

Beauty shops and perfume factories were flourishing in Egypt by 4000 B.C. The creation of cosmetics was a highly skilled and widely practiced art for men as well as women. Perfumes ("foods that reawaken the spirit," according to Mohammed) were first identified with deities; myrrh and frankincense were the ultimate of worldly tributes (the gifts to the Christ child were not casually selected). Because they served an important religious function, the strongest fragrances were reserved for the nobility. At special ceremonies and banquets, guests could appear in the presence of the pharaoh only after being sprinkled with fragrant waters. Royal mummies were deeply impregnated with lasting aromatic substances. Unguent pots in Tutankhamen's tomb, which was sealed around 1352 B.C., were still fragrant when they were opened in 1922.

During the centuries before the Christian era, every recorded culture with the exception of the Greeks adorned itself with perfumes, powders, and paint. In the ninth century B.C., during the prosperous golden age of Greek society, the dominant idea of human perfection was masculinity and natural ruggedness. Male athletes and scholars dominated the scene, and women were little more than chattel. The perfect creature was the male, unadorned and unclothed. The use of cosmetics was restricted in Greece to the courtesans, handmaidens to the noble and wealthy. These mistresses to the aristocrats were adorned with painted faces, coifed hair, and highly perfumed bodies. Added to these elaborate embellishments was a breath perfume that, in those days of uncertain dental care, may have been an essential ingredient for plying their profession. The courtesan carried an aromatic liquid or oil in her mouth. She did not swallow the breath freshener (which appears to have been history's first) but, at an appropriate moment, discreetly spat it out.

In sharp contrast to the Greeks, Roman men and women were often unrestrained in their use of cosmetics, and Roman officers often rode into battle elaborately adorned. Considerable archaeological and historical evidence exists that fashionable Roman women had vanity shelves equipped with the equivalent of every beauty preparation available at present-day cosmetic counters—and many the Food and Drug Administration would not have approved.

The Egyptians accepted the idea that the eyes, more than any other visible body part, revealed the inner emotions and thoughts of a human being. Eyes reflect an endless array of feelings, including love, hate, confusion, depression, surprise, tenderness, joy, sorrow, and despair. Quite naturally these early leaders in the art of beauty enhancement concentrated on the eyes as the focal point of facial makeup. Their favorite color for eye shadow, green, came from powdered malachite, a green copper ore mineral. It was applied heavily to the upper and lower eyelids. Then the eyes were outlined, and lashes and eyebrows darkened with a black paste made from powdered antimony, burnt almonds, a black oxide of copper, and black clay ocher.

An additional fringe benefit encouraged wide use of this mixture, called kohl. Kohl discouraged and probably killed off tiny mites whose habitat was the hair follicles around eyelashes, where they caused such assorted discomforts as itching, infections, and blindness. Many women shaved their eyebrows and applied false, expressive brows with a kohl pencil. Both Egyptian women and, later, Greek courtesans drew extended brows to meet above the nose.

Fashionable Egyptian men and women were the first to enhance their eyes with glitter. They crushed iridescent beetle shells into a powder that they mixed with the malachite eye shadow. The preferred lipstick of the time was blue-black, although Cleopatra seems to have been partial to red. Facial rouge was also red, and feet and hands were stained red-orange. Even Greek men, advocates of natural appearance, furtively added rouge to color their cheeks. The courtesans, whose goal was dramatic contrast, first coated their skin with white powder, then applied a dark red rouge.

The rouge used by civilized ancients would have failed the basic tests of the FDA. The base, made from harmless vegetable substances such as mulberry and seaweed, was safe enough. However, it was colored with cinnabar, a highly poisonous red ore mineral of mercury. Cinnabar was also the coloring material used in the lip rouge of Greek and Roman women. Doubtless much of it was ingested and in the bloodstream would have been very harmful to the fetus of any pregnant woman. Since the custom was to abandon any deformed infant at birth, there is no way of estimating how many miscarriages, stillbirths, and infants with congenital deformities resulted from these ancient beautifying practices.

The custom of staining fingernails and fingers was well established in Egypt by 3000 B.C. Nail color signified social order, the deepest red being at the top of the list. Queen Nefertiti painted the nails of her fingers and toes ruby red; Cleopatra favored deep rusty red. Women of lesser rank were permitted only pale tones, and no one dared flaunt the color worn by the queen or for that matter the king, who also displayed his superiority with brightly colored nails. Defiance of the royal color order was considered quite disrespectful and may have led to offenders being unceremoniously fed to the royal crocodiles.

Other than not infringing on the rights of royalty, few restraints were exercised against the use of cosmetic enhancement. But disapproval was voiced. Herodotus wrote that a painted prospective bride was as dishonest as a groom who overstated his property holdings. In 2nd-century Greece and 18th-century England, legislation was passed to prevent women from tricking men into marriage by the use of paints, potions, and perfumes—but such laws were short lived. Now, with a multibillion-dollar industry to support the practice, whatever sells is right!

How England's Plague Freed America

In 1665, England was struck by the Great Plague, by far the greatest outbreak of bubonic plague since the Black Death in the 14th century. England's population at the onset of the epidemic was about 460,000; by the time it was under control at least 70,000 persons were known to have died from the disease. The toll was probably much higher, as records were poorly and infrequently kept.

So many Londoners died that special arrangements had to be made for burying them. They were collected by paid corpse bearers and hurriedly interred, most still fully clothed, in great pits. The victims were left in front of their houses for the corpse collectors, just as people today leave garbage in front of their houses for collection by sanitation engineers. Londoners fled the city in droves and of course took the contagion with them, so the entire country was devastated. Eventually a great fire destroyed large populations of rats and rat fleas, and the Great Plague ended.

The Great Plague of 1665 changed the attitude of English decision-makers about emigration to the North American colonies. Before the plague, the English had perceived that the country was vastly overpopulated; there were just too many English people. The authorities' elementary but decisive solution for disposing of the surplus population was simply to ship the undesirables to America, so they began by cleaning out their prisons. Some of the first American settlers had not wanted to leave England at all, and many were marched aboard ship in handcuffs for the voyage. A few still wore handcuffs as they came ashore in the New World! The government had shipped them over to promote the formation of colonies and the foundation of new settlements.

"Undesirables" who went more willingly included religious groups such as the Puritans and purveyors of doom who saw an apparent connection between large assemblies of people and outbreaks of plague. The way these groups saw it, people who gathered to watch bear-baiting contests, theater, or other profane events were being judged by a Higher Authority when the plague descended on them. The Puritans struggled during the reign of Queen Elizabeth I to get theaters and shows closed. However, each arena or playhouse that closed for a time would eventually reopen, and many Puritans decided that relocating to the New World promised a bright new beginning.

After the Great Plague subsided, interest in relocating people to the colonies declined. Logically, since thousands of people had died in London alone, England could not possibly still be overpopulated. The shipping of English citizens to the colonies was halted, and the newly founded American colonies were virtually left to grow by themselves. As they grew, they developed into individual units where local ideas of self-government began to take shape. By the turn of the 18th century, the colonies were experimenting with the concept of freedom. The English colonies in America were ripe for independence—thanks to the Great Plague!

Laziest of All

The laziest animal species of all would have to be reptiles, and the laziest of all reptiles is the snake. People unfamiliar with or quite fearful of snakes imagine that they lie in wait under every rock or bush in the deserts and jungles of the world. These same people are usually quite surprised when, after spending days in such environments, they never see a single snake.

The snakes are there all right, but one doesn't usually see them because most of the time they are asleep or close to it. Snakes rouse into some sort of activity when hungry. After one meal they may go into a torpid state for days or weeks. When not hungry, they may stay wrapped around a branch or lie under a bush until they once again feel the pangs of hunger. During this time of inactivity a snake may remain completely motionless except to stick out its tongue to survey the environment. Snakes also rouse under other circumstances. When the urge to mate comes upon them they become instantly active!

Travelers or visitors to the desert or forest rarely see snakes because they are mainly shy of humans. More afraid of the human than the human is of it, the snake will hurry into hiding or just plain retreat rather than endure a confrontation. The only snake that would definitely announce its presence (although occasionally a little late), if approached too closely, is the rattlesnake, with its familiar buzzzzzzzz.

The Greening of the Earth

In the mid-1930s three young girls were playing in a small stream in Valley Grove, West Virginia. They were looking for something of interest, like a crawdad (crayfish), some polliwogs (tadpoles), or a horsehair snake (actually a worm, but looking for a snake seemed more fearless). If no creatures showed up they would settle for an unusual rock.

One of the girls picked up a piece of shale. The blackened imprint of a fernlike leaf clearly showed on the rock's surface; in fact it looked like part of the rock itself. Noticing that the leaf on the rock looked very similar to those on the nearby locust trees, the girls decided the imprint was made by a leaf that had fallen from one of the trees. Searching the area, they found several other rocks that displayed similar marks. Each girl took home a sample to illustrate the type of trees that grow in West Virginia. Their assumption that they had found sam-

ples of West Virginia flora was correct, but little did they realize that these trees had grown there almost 300 million years ago.

One of the girls stored the leaf impression in a drawer full of things too interesting to discard. About 15 years later she and her geologist husband, during a visit to her West Virginia home, sorted through some of her keepsakes. Her husband instantly recognized the fossil leaf imprint, so the search began for the locust trees that shaded the brook where the girls had innocently discovered a treasure of fossil leaves.

For three summer seasons her husband and several other geologists excavated this quarry. They collected several hundred specimens for the museum and laboratories of the university at which he was employed. The original discovery site yielded all its specimens during the first summer, but upstream less than a hundred yards away the scientists found an even richer quarry; some of the slabs they removed contained impressions of entire branches. Miners from the coalfields of the Ohio Valley also contributed samples that they had unearthed in their work.

The fossils were leaves from several types of trees that formed the vast coal deposits of West Virginia. This area had been a part of the great, lush swamplands that produced peat, later to be converted into coal. The leaves had fallen from the overhanging branches into the mud below, where they were buried and, over time, nothing remained but a thin film of carbon marking the shape of the original leaf. The fossil-bearing shale disappeared into a large nearby hill, carrying fossils up to the point of disappearance. If this hill could be sliced away cleanly to expose the top of the shale layer, the fossils thus uncovered would probably number in the hundreds of thousands.

The great swamp forests grew at a time when the earth was covered with vegetation. It would never again be as green. The lush greens were only occasionally interrupted by the subtle brown of dead and decaying plants. Nowhere was the color of flowering plants to be found, for they had not yet evolved. The greening of the earth was well advanced, but where, when, and with what type of plant life did it all begin?

The evolution of life took place over periods of time so great that they stagger humans' everyday time sense. It helps to imagine the history of life on earth as compressed into a 24-hour day. The first microscopic organisms originate at midnight and evolve as the day advances. Not until about 6 P.M. (three-fourths of the day already gone by) does life in the oceans become abundant. By 8 P.M. plants invade the land, and at 9 P.M. the great coal forests flourish. Dinosaurs take over the land at 10 P.M., modern flowering plants appear at about 10:30 P.M., and finally the recorded history of modern humans begins at a quarter of a second before midnight.

Three billion years ago the land was desolate and dismal, but no living things were around to lament the barren terrain. No life existed on terrestrial earth; the only sounds were the wind rushing through rock crevices and water cascading across the land, accented by explosions from erupting volcanoes that were abundant on the primeval earth. Life did exist in the sea; in fact most scientists believe that all life originated in the sea. The misty beginnings of plant life took place well over 2½ billion years ago, unceremoniously, with pond scum. Probably these earliest plants were one-celled algae that slowly evolved into complex, multicelled plants.

Gradually, about 420 million years ago, the first pioneering plants moved out of the swampy areas and began a life on the barren world of naked rocks unsoftened by a green mantle. No scientist can say which plants made the first hesitant steps on land. They were undoubtedly borderline forms that developed from waterweeds in coastal swamps and still lived a half-water, half-land existence. Compared with today's complex and varied plant life, this early vegetation would seem insignificant. The plants lacked true roots and leaves and, deprived of continuous contact with moisture, died quickly. Their foothold on land was nevertheless secure; it was a staggering event for the history of the earth.

By the mid-Devonian, 375 million years ago, a great surge of vegetation swept the earth, producing a landscape that resembled the swampy Florida Everglades. Land plants, very scarce before Devonian times, diversified greatly within 30 million years. A sampling of a mid-Devonian landscape can be seen in an exposure of rocks near Gilboa, New York. It includes petrified stumps of trees 3 feet in diameter, estimated at over 40 feet tall, and slender 100-foot ancestors of the modern club mosses, which are less than 12 inches high. These are the fossil remains of the earliest known forest that ever clothed the land.

But many things were missing from this dawn forest. It was still an almost soundless world, for animal life on land was a rarity. There was no buzzing of insects, singing of birds, roar or bellow of mammals, not even the hiss of a snake. Animal life included a few mites and several forerunners of insects and spiders that had crawled ashore. This was the vanguard of a host that would soon populate the earth.

The world of 300 million years ago was springtime in the history of plants, an unbelievable flourishing and harvest of plenty. In this, the Carboniferous Period, warm tropical swamplands prevailed over much of the world. Dense jungles housed giant ancestors of modern horsetail, some growing almost 130 feet tall, surrounded by 40-foot fernlike trees. With plentiful swamps and forests, crawling and flying insects diversified greatly. Without flowering plants, however, there was little use for bees. The land was still the many shades of green.

Scientists refer to this era as the age of cockroaches, an accurate description because cockroaches were quite abundant. They were similar to today's cockroaches, except in size; some of them were over nine inches long. Overhead flew the ancestor of the modern dragonfly, with a wingspan of nearly three feet, that easily preyed on the cockroach.

Another denizen of the lush forests was a centipede over six feet long. Primitive amphibians and reptiles also populated the earth, but no mammals had yet evolved.

In the coalfields of South Joggins, Nova Scotia, fossil tree trunks have been found buried in shale. Amazingly many of them are still standing upright on the very spot where they grew some 300 million years ago! Most of the preserved trunks range from one to three feet in diameter, and for some of them as much as nine feet of trunk has been preserved. Judging from the diameters, some of the trees must have reached gigantic heights.

The trunks were preserved by the accumulation of coal materials and clays that slowly buried them. As decay set into the exposed dead wood, only the buried portion of the trunk remained. The interior of each standing trunk was also completely decayed before final burial, leaving a great hollow interior. Actually, only the outer rim, including the bark of the tree, has been preserved.

The hollow interiors of the trees did not remain empty for long. As the standing trunk was slowly being covered by accumulating sediments, its summit must have for a time been level with the accumulating soil. The hollow thus served as a well, or pit, for snails and millipedes to crawl into. Many of them died and were buried by the clay material filling this hollow interior.

These pits in the soil were probably concealed by fallen leaves from nearby living trees. Many of the early land vertebrates, particularly amphibians, fell through the soft cover, unmindful of the trap that was concealed under the leaves. With a cavity too deep to climb out of, the hollow tree became a veritable death trap. From only 25 cataloged trees, over 53 individual specimens representing 12 different species have been recovered. The character of the sediments in the trunks changes often, indicating that filling the hollow trees was a very slow process that took a great period of time.

No doubt today's immense coal deposits were formed when these giant forests were buried where they grew. Coal has been formed in small quantities ever since vegetation began to grow on the land, although plants were not abundant enough before the Carboniferous Period to produce the deep coal beds mined today. The luxuriant forests of the Carboniferous are the source of almost all the coal that has fueled modern industry.

The conversion of plants into coal occurs when vegetation does not decay. This happens when the forest becomes a swamp of standing water; falling trees sink into mud that lacks enough oxygen for the agents of decay to rot them. Instead the plants slowly begin to turn into peat, the first stage of coal. Peat can lie in bogs for thousands of years without changing much. But in time sediments carried into the swamps by rivers bury peat deeper and deeper, compressing it into lignite, a low-grade coal. Under the weight of millions of years of additional deposits the lignite is further compacted into bituminous, or soft, coal. Finally additional pressure brought about by shifts in the earth's crust compresses the material in many places into anthracite, or hard coal.

The amount of compression needed in the conversion of coal is tremendous, for one foot of bituminous coal represents about 20 feet of original plant matter. The extent and luxuriance of these Carboniferous forests are demonstrated by coal seams upward of 400 feet thick in China. This represents about 8,000 feet of original vegetation. In Pennsylvania, Ohio, and West Virginia the coal seams are only 5 or 6 feet thick on average, but they blanket thousands of square miles. Apparently, widespread forests flourished on top of previous forests, and each in its turn became compacted into a layer of coal. One West Virginia coal bed has yielded 120 such seams, one above the other. The swampy environments were not continuous in time; they alternated with dryer conditions. But when they did occur they were always very widespread. Geology textbooks rightly refer to the Carboniferous Period as the age of coal.

The giant coal-yielding forests came to an end about 230 million years ago. Vast changes on the earth's surface resulted in mountains, glaciers, and deserts quickly wiping out the expanded swampy environments, and vegetation took on a more modern aspect. A new era began—the age of dinosaurs—but the overall green color persisted. About 100 million years ago, during the last period of the dinosaur era, the Cretaceous, the green blanket that covered the earth began to change. Flowering plants arrived at last, as did pollinators such as the bee and the hummingbird. Green was now one of many colors that blanketed the earth.

Peter's Mistake

Peter I (1672–1725), czar and emperor, was one of Russia's most energetic, talented, and ruthless rulers. Proclaimed "Peter the Great" in his own time, he was a ruler of truly grandiose vision and personality and intimidating energy. Through his reign he was animated by a sense of duty that he could never instill in the majority of his subjects. This showed up in one of his many tours and journeys.

At heart Peter the Great was a naturalist, collecting biological and mineral specimens wherever he went. On one of his tours he collected well over 1,300 specimens, which he preserved in brandy. That turned out to be a huge mistake. During shipment of the collection back to Russia the crew discovered the preservative. Before the ship docked the sailors aboard the cargo ship drank all the brandy in which the specimens were preserved. What arrived in St. Petersburg was a colossal mess.

What happened to the sailors involved in the drunken sea voyage is not recorded. It is doubtful, however, that the "Great One" merely chalked it up to a learning experience.

Venus

"We can learn even from nightmares."

Sigmund Freud

Earth is about 93 million miles from the sun; approximately 26 million miles nearer to the sun is the planet Venus. Originally named for the Roman goddess of love and beauty, Venus was until quite recently identified principally as the morning and evening star because it is brilliant enough to be seen brightly even when the sky is still partially lit by the sun. Being closer to Earth than any other planet, Venus naturally appears very bright, but the dense cover of white clouds that cloaks the planet and glares in the light of the sun contributes more to its brilliance.

When Galileo first gazed at Venus with his newly devised telescope he could tell it was a sphere because it waxed and waned in phases, from crescent to a full circle, just as the moon did. He could see nothing of its surface features, however, and for the next three centuries no astronomer saw any more than Galileo did. Science fiction writers were free to dream up Venus as an imaginary paradise. Astrophysicists could reason, with merciless logic, that the oceans and atmosphere of Venus were laced with valuable oils and that this was a watery, very livable planet. As long as the cloud cover kept the surface a mystery, even researchers equipped with the most powerful telescopes were forced to guess and imagine.

Venus is so similar to Earth in size and weight, and in its chemical elemental makeup that referring to the planet as Earth's twin was quite natural. A pronounced difference between Venus and Earth is their rotation. Venus takes 243 earth days to complete one rotation (one "day"), which is remarkably sluggish for a planet. Also, its rotation is retrograde, with the sun rising in the west and setting in the east. Scientists believe that Venus may have had a normal rotation during its primeval days but some eons ago suffered a titanic collision with another celestial body that sent the planet reeling backward. Oddly, the outer atmosphere of Venus is moving very slowly; it is literally wrapped in a violent whirlwind! Venus takes 225 earth days to make one complete revolution around the sun; Earth, of course, completes a longer annual trip in 365 days. A Venus day is longer than its year!

Venus's reputation as a planet of beauty and mystery remained untarnished until the last few decades. The Soviets made several successful Venus probes in the early 1970s. Their earlier attempts had been unexplainable failures; perfectly functioning equipment stopped transmitting long before it reached the Venusian surface. Finally the Soviet scientists arrived at the inescapable conclusion that the probes had been crushed and/or burned in Venus's atmosphere.

Then in 1978, the United States was successful with the *Pioneer Venus* craft, which has been circling the planet continuously and mapping its surface. Soviet and U.S. probes have found the terrain to be remarkably bare of dust or soil and covered with basaltic rocks that resemble the lava flows on Earth's ocean floors. No doubt these rocks are basaltic flows kept smooth and flat by the immense pressure of the atmosphere, just as oceanic water pressure keeps the marine lava flows flat-faced and spread out.

Scientists now know that Venus is a waterless and hostile world, baking in a hellish heat of about the temperature of molten lead, 900 degrees Fahrenheit. It is oppressed by a thick atmosphere of carbon dioxide and sulfuric acid clouds that bear down on its surface with a pressure 90 times greater than that of Earth. This is equivalent to pressure 3,000 feet below the surface of the oceans. Only about 2 percent of the solar energy penetrates the thick cloud cover and reaches the surface of the planet. Clearly the thermal energy that does reach the surface is trapped by the carbon dioxide atmosphere and the clouds of sulfuric acid. This produces an exaggerated form of greenhouse effect. Continuous for billions of years, the effect has made the surface of Venus into a superfurnace that would easily melt lead.

Recent Venus probes have indicated that the planet's atmosphere is more dynamic and complex than that of Earth. Instruments have recorded frequent lightning discharges with thunder that reverberates for 15 minutes or more! On Earth neither lightning nor thunder lasts more than a few seconds. Its existence on Venus indicates that the range of conditions under which lightning can occur is much greater than previously imagined. The bolts of lightning must be incredibly large, and the frequency suggests that the surface of the planet is never really blacked out but is constantly bathed in a flashing eerie greenish light.

Scientists believe that acid rain falls almost constantly from the thick cloud covering. When it gets close to the planet's surface, however, the intense heat separates the precipitation into its component parts; none of it reaches the surface of the planet. On Earth such a phenomenon, called "ghost rain," occurs in hot deserts such as Death Valley. The rain falls from the clouds in heavy loads, but heat causes it to evaporate before it can reach the desert floor. So it returns to the clouds from which it came. Just as on Venus, scarcely a drop of the rain hits the ground below.

On May 4, 1989, the unmanned spacecraft *Magellan* was launched from the shuttle *Discovery*. The *Magellan* is equipped with an imaging radar system designed to "see" through Venus's cloud cover and obtain detailed photograph-like images of the planet's surface. It reached Venus on August 10, 1990, and began taking the sharpest images of the planet ever seen. This spacecraft was placed in an elliptical orbit, and about 90 percent of the surface of the planet has been radar photographed and mapped to date.

Like a crazed plastic surgeon, plate tectonics continually reshapes the face of the earth, tearing the planetary skin in some places and tucking it away in others. Images of Venus sent back by the *Magellan* probes suggest that certain elements of plate tectonics have also scarred this nearby planet, even though the mechanism appears somewhat different. The surface of Venus contains the same type of geologic features found on Earth, and they can best be explained by plate tectonics. Deep valleys, huge volcanoes, extensive plateaus, and hills are present on both planets although they differ in size, shape and distribution. A major difference between the two planets is that the surface of Venus is mostly flat, gentle rolling plains.

An outstanding geologic feature reported by radar findings is a canyon on Venus that is larger than Valles Marineres, present on the planet Mars. The latter was previously the largest known canyon in the solar system. Mapped by radar, the Venusian canyon's measurements are 3 miles deep, 175 miles wide, and over 900 miles long.

Scientists now know that Venus is a far cry from the moist, life-giving paradise once envisioned for the planet of love and beauty. In fact, of all the planets in our solar system, Venus is the closest to what Dante must have envisioned for his inferno. Getting right down to it, any astronaut to land on this planet and step out of his or her protective spacecraft would be instantly poisoned, squashed, and fried.

Stone Age Rolling Stones

In 1989 a group of Russian anthropologists assembled a band of prehistoric musical instruments unearthed in various caves and other encampments. The instruments, percussion types, looked like anything but a symphony orchestra. Most consisted of hollowed bones and reinforced mammoth skulls. Doubtless the early musicians had also used hollowed-out logs, but these were unable to endure through hundreds of centuries.

One can almost picture these Stone Age musicians beating out a concert. Saturday night, 25,000 years ago, must have scared the dickens out of roving saber-toothed cats. They probably thought the heavy booming was made by some huge animal pounding on the ground. Time to seek out better hunting grounds.

For a long time scientists believed the earliest musical instruments were some sorts of drum, such as those the early Russian cave band used. But recently in the Haua Fteah Cave in Libya, scientists uncovered several fossil bones carved into whistles. These artifacts, used by Neanderthal musicians, are the oldest known musical instruments.

Life was harsh in those days, and a little music must have been very soothing to the savage soul. Scientists believe these early whistles were played occasionally for pleasure. Doubtless, however, they were used mostly for imitating the calls of birds and animals to lure them toward the hunter. The shrill notes may have been designed to drive neighbors into a frenzy, since the music produced wasn't exactly a Mozart musical masterpiece. A Stone Age musician could have been discouraged from furthering a musical career by getting his head bashed in by a neighbor crazed by the headache that high shrill notes can cause.

CHAPTER SIX

Waiting for the Big One

The term *terra firma* ("solid earth") is comforting but a blatant misnomer because the earth's crust is truly dynamic and almost constantly in motion. This is evidenced by the occurrence of more than a million earthquakes per year throughout the world. An earthquake can be felt somewhere in the world every few minutes, but happily for most creatures and things only 1 in 50 causes damage.

The Richter scale, which measures the ground motion as recorded on seismographs, charts the magnitude of every quake. It was originally designated as a logarithmic scale to classify the amount of energy released in an earthquake. Calculated in megatons (one megaton is equivalent to a million tons of TNT), it is an open-ended scale, from one through nine-plus, that compares the amount of energy released to an equal number of hydrogen bombs. The 1906 San Francisco earthquake, estimated at 8.3, produced the same energy as several thousand atomic explosions.

California is indeed earthquake country; the state experiences around 10,000 quakes per year. Most are imperceptible to humans, but Californians do feel about 500 annually. Fewer than one per year causes any damage or death, partly because of rigid building codes throughout the state. Most California quakes are generated by the famous San Andreas Fault that runs through southern California for a distance of 650 miles and cuts through the earth's crust about 20 to 30 miles deep.

The San Andreas is the hinge line between the American Plate on the east and the Pacific Plate on the west. It seems to be acting out the Old Testament prophecy of Zechariah that states, "The Mount of Olives shall cleave in the midst toward the east and west, and half the mountain shall move toward the north and half to the south." The plates slide horizontally past each other at the rate of two to three inches per year. This seems to describe a strike-slip motion along a fault as the Pacific Plate moves to the north and the American Plate to the south. This movement along the San Andreas tends to stretch and squeeze the adjacent land which, to relieve the stresses placed on it, will fracture and move. The pressure is transferred to other sections of the earth's crust, and they will also fracture. It appears to be an almost endless process.

The southern extension of the San Andreas underlies the Gulf of California; the northern extension enters the San Francisco Bay and terminates against the Juan de Fuca Plate near Cape Mendocino in northern California. The American northwest is relatively free from the activity on the San Andreas Fault. The Juan de Fuca Plate is a more menacing culprit.

At 5:34 A.M. on March 26, 1993, Oregon was shaken by a 5.3 earthquake that rattled much of the state and caused considerable damage, although fortunately no deaths. This was a rude awakening for Oregonians, who think of earthquakes as a California anomaly. In California, a Richter scale reading of 5.3 is not even considered a moderate quake, although it can cause damage to old, unreinforced structures. This is what occurred in Oregon on March 26. Oregonians may be surprised to realize that they undergo about a thousand earthquakes per year, most of such low energy levels that they go unnoticed.

This situation is most likely going to change. A team of scientists from Washington State has found disturbing evidence that the Pacific Northwest has experienced several powerful earthquakes in the recent geologic past and may be headed for another catastrophic shock in the not too distant future. Their concern is even more intense because northwestern cities such as Vancouver, Seattle, and Portland are not prepared for high-magnitude quakes. Should a trembler of 8+ on the Richter scale strike these areas, the destruction could exceed the damage done in the San Francisco disaster of 1906.

If, or when, a giant quake rattles the Pacific Northwest, the rock slippage will originate 10 or more miles beneath the surface from a tectonic structure called the Cascadia subduction zone. The zone runs offshore from Vancouver Island to Cape Mendocino and marks the place where a piece of ocean floor, the small Juan de Fuca Plate, is slowly crashing into the edge of the North American Plate. As they undergo collision, the lighter and more buoyant North American Plate runs over the Juan de Fuca, pushing it down into the earth's interior. Because of friction with the North American Plate, some of the Juan de Fuca undergoes melting. The molten material finds its way to the surface through cracks and fissures and produces the series of active volcanoes such as Mount St. Helens.

Earthquakes originating in subductive zones are fairly common. They are often of great magnitude, such as the 1960 quake in Chile that reached 8.9 on the Richter scale. Even more unforgettable was the Alaskan quake of 1964, which reached a Richter high of 9.1. It released 12,000 times more energy than was released by the atom bomb dropped on Hiroshima. Such massive jolts occur typically when the subducting plate fails to slide smoothly under the overriding plate. When two plates lock together, as they do from time to time, they will build up strain for hundreds of years. The pressure responsible for this stress continues to thrust and shove on the plate. When the strain builds to a point where the locked zone can no longer resist the stress, the plates suddenly slip past each other, generating a massive earth shudder.

Concern is growing about the Cascadia subduction zone. Sensitive seismic instruments have not detected any large tremors originating from the contact of the North American and Juan de Fuca Plates. This could indicate that the plates are tending to stick and are therefore preparing to unleash a killer quake. Historic records are of little use, since they go back only to the early 1800s and since that time no large earthquakes have struck the northwestern part of the United States.

For earlier evidence the scientists have investigated the geologic record. In 1987 they reported finding strong evidence that sections of the Washington coast subsided at least six times in the last several thousand years. The scientists have traced the evidence up and down the coast and, by using radiometric carbon-14 dating, have established the timing of these events. These instances of sudden subsidence or uplift have been traced as far down the coast as northern California.

Although land can rise and subside for reasons other than earthquakes, the evidence here leaves little doubt as to the quake origins. Mud that had collected in estuaries was found in the sediments deposited directly on top of dry land soil layers. This suggests that the lowland areas abruptly dropped below the high-tide levels and were quickly covered by a deposit of marine mud. If the lowland area had subsided slowly, there would be a gradual transitional zone between the terrestrial soil beds and the overlying marine mud instead of the sharp distinction between the two.

In these same sedimentary deposits were valid signs of tidal waves (tsunami) that coastal earthquakes can generate. Some of the sediments that subsided show a sheet of coarse sand packed between soil and mud layers. The sandwiched sand layers reflect a series of enormous quake-generated tsunami that crashed over the subsided sections of coastline. The tidal waves deposited the sand, which was immediately covered by fine-grained marine mud. Research geologists examined the sedimentary deposits still visible on the Chilean coast after the 1960 quake. Comparing them with the northwestern sediments yielded similar examples of subsided coastline and tsunami deposits of mud.

Some of the most dramatic evidence of prehistoric quakes in the Pacific Northwest is found in the numerous trunks of red cedar trees that have remained standing centuries after they perished. Tidal muds buried the lower trunks, suggesting that the ground level dropped below the high-tide mark, and saltwater flooded their root systems. Tree ring dating established that the red cedar trees died somewhere between 1684 and 1687, all at the same time! This implies that a section of the Washington coast over 90 kilometers (56 miles) long dropped suddenly below sea level about three centuries ago. Tree ring evidence coordinates quite accurately with the data provided by radiometric carbon-14 dating.

Based on the combined evidence, the conclusion is that most of the coastal areas along the subductive zone suddenly subsided or uplifted during the late 1600s. Unfortunately, the Cascadia quakes do not hit with predictable frequency. Although the events average about 600 years apart, the span between quakes has been as short as three centuries. Moreover, these coastal areas have been subjected to this type of tectonic plate disaster for at least the last 10,000 years.

Accumulated evidence seems to indicate that the Cascadia subductive zone is storing up energy for a future blockbuster quake. Data collected over the last 60 years have shown that the coastline bordering the subductive zone has undergone significant warping during that time. Apparently this zone is at least partially locked and is building up an incredible amount of strain as it bends the edge of the North American Plate. When the plate has had all the stress bending it can take, it will snap back wildly. When this happens, Portland or Seattle may become the first city on the moon!

Phantom Treasure Ship

Phantom ships, such as the classic *Flying Dutchman*, are the substance of many legends. The ghost ship, because of some curse or as retribution for some unpardonable crime, is unable to reach any port and must wander endlessly over storm-tossed seas. Not all of these ships have sprung from the storyteller's imagination. A more recent story about a phantom ship, the freighter *Baychinco*, is based on recorded fact.

The *Baychinco* became stuck in the ice off Point Barrow, the northernmost promontory of Alaska, in the year 1931.

Freighters have often become immobilized by encroaching Arctic ice, and the crew met the emergency in a most reasonable and prudent way. They remained with the ship to await a breakup of the ice. When the area was hit by a sudden severe blizzard, the ship rocked so violently that the crew feared that it would break up. So the captain ordered all aboard to abandon ship. The crew descended onto the thick ice alongside the imprisoned ship and lay flat on the solid ice sheet for protection from the storm.

Finally the storm abated, and the air cleared enough for them to see around them. To their shocked amazement the freighter was gone without a trace, with no debris or any other evidence that it had broken up and sunk during the storm. The only explanation was that the intense winds had broken the ship loose from the confining ice and, with the help of the wind, it had drifted free. To the bewildered and stranded crew the ship had simply vanished from their sight, forever.

The *Baychinco* became one of those spectral ships that loom indistinctly on the horizon, enshrouded by fog, snow, squalls, or other tempestuous weather. It was a true "ghost ship" that continued to drift aimlessly among the Arctic ice packs. For a number of years the ship was much sought after because it carried a heavy cargo of valuable furs. It has been seen several times, but ice packs have always prevented any ship from making contact. The *Baychinco* was sighted and identified at a distance as recently as 1964.

An Invasion of Guam

In June 1944 a platoon of U.S. infantry soldiers was on combat patrol in Dutch New Guinea (now West Irian). As they pushed through dense underbrush one of the soldiers reached up to balance himself by grabbing a low-hanging branch. This could have been a fatal move. The branch was already occupied by a five-foot brown snake that promptly bit him on the thumb. The two soldiers standing behind him immediately reacted by attacking the snake with unsheathed bayonets and hacking it into oblivion. They dared not fire their rifles because the enemy would certainly hear the shots and set up an ambush.

The bitten soldier sank to his knees, his face ashen with fear. The platoon medic hurried to the scene; he was a medical doctor who by coincidence knew something about snakes. He examined the remains of the brown snake and eased the victim's mind by identifying it as a species of *Boiga* that was only mildly poisonous. Apparently he was right, for the soldier was sent to a rear base hospital for only a few days and released back to duty.

Specimens of this variety of snake have continued to bite creatures, including people, not only in New Guinea but also on the island of Guam. Properly identified as a tree snake, *Boiga irregularis* is a natural inhabitant of New Guinea and Australia. So what the heck is it doing on Guam, over 1,200 miles to the north?

Guam is the largest and southernmost of the Mariana Islands in the west central Pacific. An unincorporated territory of the United States and an important military base, the island is only 30 miles long and from 4 to 8 miles wide. With a total area of about 209 square miles, Guam is smaller than Molokai, fifth in size of the eight Hawaiian Islands. Bats, rodents, lizards, and several birds are the principal native animals, but deer and wild pigs have been introduced. Into this island paradise came the brown tree snake, *Boiga irregularis*.

A rear-fanged snake of the subfamily Borginae, the brown tree snake has a couple of enlarged teeth in the back of the upper jaw that are grooved from top to bottom. The teeth underlie the poison gland so that, when the snake bites, the gland secretes a mild venom that trickles down the tooth grooves into the victim's wound. The snake is up to 10 feet long, but it has a relatively small head, so the mouth doesn't open wide enough to get a good grip on its victim. This, along with the mild venom, may explain why there have been no human casualties even though a number of residents have been bitten.

The snake appears to be mainly nocturnal, although many are seen during the day. *B. irregularis* climbs trees with great speed, and it is also quite agile on the ground, where it almost always seems to travel in a straight line. It invariably seeks to reach the nearest high point, be it a tree or telephone pole, and its main prey is definitely birds!

Prior to the 1950s, Guam had six endemic species of birds (they existed nowhere else in the world) and several species of native, nonendemic, birds. Almost all are now extinct on Guam, along with three species of bats and a variety of lizards. The decline of birds was first noticed in 1978, but its cause was not identified until several years later, and the situation continued to worsen daily.

About 1952, a military cargo plane landed in Guam after a brief stop in New Guinea. The supplies were unloaded, along with an unobtrusive stowaway, a pregnant *Boiga irregularis*. With no apparent threats in her new home, she produced her brood in safety. To the brown tree snake this was a veritable Garden of Eden. With little competition and no natural enemies, *B. irregularis* multiplied completely out of control.

In less than 40 years the snake population that began with a single pregnant female has exploded to an estimated one million to three million. This averages out to about 10,000 snakes for each of Guam's 209 square miles! Since snakes don't space themselves evenly, their density in one particular area was calculated in 1992 at 30,000 per square mile. That would amount to 62 snakes per football field heading for the goalposts and light poles and hundreds of others in bleachers, press boxes, and control booths.

In 1982 concerned citizens of Guam brought in a biologist to find what was causing the decline in their bird population. Eventually the biologist turned her attention to the rather plentiful snake that natives called the Philippine rat snake. The first one she cut open had birds and eggs in its digestive tract. She offered a bounty for each "Philippine rat snake," and virtually every one of the throngs of snakes brought to her laboratory showed remains of birds and eggs. The culprit was *B. irregularis*, which continued to multiply as the birds declined.

Before the brown tree snake arrived from New Guinea, Guam had been almost as snakeless as Ireland. The only native serpent was a burrowing blind snake that looked like a worm and fed on termites. The birds of Guam evolved in the absence of snake predators, so they were naive, incautious, and totally defenseless in the face of an enemy. *B. irregularis*, having left its enemies behind, faced a bountiful food supply on a utopian island, but only until it completes the job of exterminating all edible animal life. Most of the surviving birds of the remaining endemic species have been taken from Guam to zoos on the U.S. mainland. The goal is to breed them back into populations large enough to be released back on Guam; in the meantime *B. irregularis* must be controlled or eliminated from the island.

At present, people who go for a nature walk or drive on Guam see snakes draped over power lines, telephone poles, and just about every tree around. Electrical power outages are common as snakes slither into power boxes and cause short circuits. People have become accustomed to being plunged into darkness; over 50 outages a year occurred during the 1980s. But herein may be a partial solution to the overabundance of *B. irregularis*.

Recently two repairmen investigating an electrical short circuit found, not to their surprise, a well-toasted Boiga coiled around the power box. They took it home and, with a courage based on "don't knock it till you've tried it," sampled some of the fried snake. They were amazed to find that it tasted like well-seasoned chicken, and it proved to be quite digestible. The word spread about the reptilian

chicken, and others who dined on this inexpensive substitute found it entirely acceptable. Perhaps the brown tree snake has finally met its predatory match, and humans will be able to partially undo what they were inadvertently but carelessly responsible for doing. If Guamanians serve enough snakeburgers in place of hamburgers, the ecological system of the island may self-correct. Then, for the remaining birds, bats, and lizards, paradise may be regained.

However, it is most unlikely that human predation will keep up with the population explosion of *B. irregularis*, so wildlife specialists on Guam are advising folks to learn to live with snakes. This includes accepting their invading homes, even attacking infants, and being a part of all outdoor activities. Unfortunately, their original invasion was so easily accomplished that it could be repeated. The brown tree snake, *Boiga irregularis*, has been found on aircraft that have arrived on Hawaii from Guam. Another paradise lost?

How to Fix a Broken Arm

To anyone working or playing in the desert, a welcome sight is the cottonwood tree, the shade tree of the plains, prairies, and deserts. No respectable ranch anywhere at any time would be complete without the cottonwood.

As important as its shade might be, other uses for the cottonwood were discovered during the early days of taming the West. Both animals and humans were aware of its value as an indicator of moisture somewhere below the surface. Native Americans and early settlers soon discovered its therapeutic value in setting broken bones. By boiling down cottonwood bark, they obtained a thick, honeylike syrup that would set broken bones in a remarkable cast.

The bark was placed in a large vessel and hung over a fire for cooking. The process was not lengthy, but to the person with the broken limb who had to wait until the bark was collected and the syrup prepared, it may have seemed interminable. The injured arm or leg was wrapped in a cloth, and the cooled syrup was spread completely over the wrapping. This quickly hardened into an effective immobilizing cast. The cast disintegrated in about two months, just long enough for the bone to heal.

Improved Birdbrains

The term *birdbrain* is commonly used to describe a nitwit, a stupid or silly person. Birds earned this reputation because some exhibit little intelligence and engage in life-threatening behaviors that are downright stupid. Woodpeckers will hammer holes in a tree and stuff in acorns and nuts for the winter. They don't notice when the tree is quite thin or is the wall of a cabin, so hundreds of food bits fall to the ground or indoors, inaccessible to the hungry bird. The woodpecker may be upstaged by the hoatzin of the Amazon valley, which builds its nest on low branches overhanging the water. At the first sign of an intruder the hoatzin dives to the safety of the water but seems unable to understand that the intruder may be a crocodile resting directly under its nest with jaws agape.

Among species of birds that are downright intelligent are the 105 members of the Corvidae family, which includes jays, magpies, ravens, and crows. The large, noisy, mischievous common crow appears to be the Phi Beta Kappa of the feathered world. Of all bird brains that have been studied, the crow has the largest cerebral hemispheres relative to body size. This denotes an excess of brain power and explains why crows can learn even faster than monkeys. They can be taught to read clocks and to count, and they are seldom fooled a second time by a scarecrow.

Crow ingenuity enables them to flourish despite incessant attempts to eliminate them, for they are something of a pest. Clearly the crow is a nuisance to a number of animals, as it amuses itself by tweaking tails of birds, pecking at sleeping cows and dogs, and engaging in other annoying but harmless fun. Some of the crow's devices involve intricate problem solving. In Finland, people fishing through holes in the ice often return to their line to find a crow pulling it in with its bill. Then the crow returns to the hole, walking on the line to keep it from sliding, and retrieves the fish at the end of the line.

In recent years crows have been flourishing as urbanites, especially since their discovery of the four-wheeled nutcracker. Any patient birdwatcher can observe crows sitting on a tree branch near a road and waiting for an oncoming car or truck. Just as a car approaches, they will fly over the road and drop nuts in front of the vehicle. The car will, of course, run over the nuts, thoroughly cracking the shells. The bird then dines on the exposed nut meats. Such meals are not without hazard. If the road happens to be a busy freeway, the crow must grab and fly. The clever crow that indulges in a leisurely lunch often becomes roadkill.

In the city parks of Japan the green-backed herons obtain their food without exposing themselves to danger. They drop small twigs into a pond and watch as the twigs float a short distance. They drop a few more until one of the floating twigs attracts the attention of the small minnows that inhabit the pond. Thinking it may be something to eat, the fish swim up to the twig. Their investigation doesn't last long because the heron quickly snaps them up when they come into range.

The tropical bee-eater puts itself in jeopardy each time it enjoys its one-course meal. But it has perfected a ritual that renders the bee a juicy, nonthreatening morsel. The bee-eater catches a bee in midair by grasping it around the waist with the tip of its beak. Then it bangs the bee's head several times against a nearby hard object and in the process shifts its grip to just in front of the stinger. The bird rubs the stinger vigorously against a solid object until it breaks off or the venom escapes. Finally the bee is almost suitable for eating, so the bird juggles it back to its original position, bangs its head a few more times to make sure it's dead, and gulps it down.

A Walk in the Sun

During the late 1980s a group of scientists from the University of California, Riverside, unearthed a series of footprints made by ancient Native Americans wandering along the banks of the Mojave River in San Bernardino County, California. These are the oldest known human footprints in North America.

Judging from the size of the prints, these early wanderers comprised two adults, a male and a female, and two young children—just a typical American family out for a walk in the sun. The scientists found a total of 54 prints. The adults were walking in a southerly direction, and the children appear to have been playing, as their tracks go in many directions, although they generally followed their parents. On the soft, rain-soaked ground the tracks left deep impressions as evidence that people had passed by.

A short time, possibly a week, after their passage a grass fire swept the area and baked the clay, thus hardening and preserving the footprints. Charcoal from the burned grass, subjected to radiometric dating, yielded an age of over 5,000 years.

The forces of erosion and weathering usually obliterate signs such as tracks, so a find of this sort is considered quite spectacular. Nevertheless, a fortunate set of circumstances does occasionally occur, and a few of the trillions of tracks left by humans and other animals have been preserved. In Nicaragua a series of slightly older human footprints has been preserved for about 6,000 years since several individuals walked over a soft volcanic mud one day. A volcano had evidently erupted, and a shower of ash covered the area. Rain turned the surface into soft mud, and over this surface the ancient people walked and left their marks. Their flat prints close together suggest that they were in no hurry, and all were walking in the same northerly direction. The hot tropical sun baked and hardened the volcanic mud, retaining marks of the footprints. Another eruption followed and ash buried the prints, thus guaranteeing their preservation.

The study of fossil human footprints has become of primary interest to paleoanthropologists. By determining where, when, and with

whom the early humans walked, scholars can make many inferences about lifestyle. The prints found in the cave of Le Tuc d'Audoubert reveal much about the customs and religion of the Cro-Magnon. The cave's large galleries and chambers have remained intact and almost completely unchanged since Paleolithic times. The visitor proceeding down the narrow passage will pass several prehistoric footprints of humans and cave bears. Finally, over 2,000 feet from the entrance, is the place where the famous Bison Sculptures are located. The clay bisons, a male and a female, are each about two feet long. These sculptures are no mere primitive carvings, but are works of surprising plastic beauty, rich with life and filled with an astounding force of expression.

The floors are covered with footprints of the artist. The oversize prints indicate that he must have been a large man, and this is confirmed by the imprints of his broad, heavy fingers on the clay as he molded the bison. The main evidence of paleolithic activities at Le Tuc is documented on the smooth floor of a low-roofed chamber. Some extremely simple abstract finger tracings are etched on the floor, along with 50 deep human heel marks visible in the clay.

The absence of any toe prints suggests that whoever walked across the chamber did so in a most unusual fashion. All of the heel prints are small and narrow, evidence that the room was used by children or adolescents. This is further confirmed by the fact that at no point is the ceiling higher than five feet. The deeply impressed heel marks validate the conjecture that this chamber was used for the ritual initiation of young hunters. The heel marks follow five different paths, indicating the ceremony was directed and controlled as a sort of ritual dance. Further evidence of a puberty rite is borne out by the presence of five sausage-shaped clay forms, obvious phallic symbols. Since the room was well away from the normal route through the caves, its use was apparently restricted to initiation ceremonies practiced so frequently among primitive peoples in the modern as well as the prehistoric world.

Such exceptional remains as artwork and footprints are most likely to survive only in very remote chambers of a cavern, where they are protected against vandalism and casual explorers. Even more important, conditions of temperature and humidity must remain fairly constant regardless of weather changes in the outside world. Le Tuc d'Audoubert is one such preserve; another is the painted cave of Niaux in the central Pyrenees.

In a newly discovered gallery at Niaux are some 500 well-preserved human footprints that seem to be associated with five black animal paintings—a horse, three bison, and what appears to be a weasel—grouped in one part of the gallery. The footprints are of individuals of all ages, and they seem to be performing a ritualistic dance in front of the paintings. The children's prints suggest that they did not take the services very seriously. They seem often to be at play, chasing each other or dabbling in the mud, making impressions of their hands or other childlike images. The painted chamber is at least 2,000 feet from an entrance to the cave, so these remote salons were obviously for ceremonial use and not day-to-day living.

To stand silently in this painted chamber covered with the footprints of antediluvian life is a strange experience. Visitors have said they felt the presence of those ancient hunters so vividly that many succumb to the temptation to look over their shoulder to be sure that no one else is standing in the chamber. Nearly 200 centuries separate the visitor to Niaux from the children who danced and chanted their way around these muddy galleries. Since they left 20,000 years ago, their ancient playground has remained undisturbed until the present, except for one incident.

Three young visitors entered Niaux from an unidentified entrance long after the ice age hunters had passed into oblivion. These later visitors appear to have been three children who carried wooden torches and left small footprints on a sand bank near some of the wall paintings. They must have felt like visitors to the Louvre in Paris as they viewed the ancient art. Burned carbon droppings from their torches permitted scientists to apply radiometric dating to the time of their visit. It was calculated to be about 8000 B.C., over 10,000 years after the Stone Age paintings were executed. Apparently no other human was in this gallery at Niaux until its discovery in 1970.

The Cave of the Witches, near the city of Genoa in northwestern Italy, was so named by the ancient Romans because they believed it was inhabited by witches. Many years before the Romans, however, the Stone Age Neanderthals had lived there and had their own superstitions about the cave.

Evidence of occupation by Neanderthals had long been established, but scientists believed it to be more than a residential site. In 1950 scientists opened a new section of the cave, revealing a large chamber in which beautiful stalactites hung from the ceiling and impressive stalagmites grew from the floor. The entrance to this chamber appeared to have been blocked by a natural landslide. This was not the case, however. Eons ago powerful arms piled huge rocks at the opening, sealing off the interior. With the passage of time the barrier took on the aspects of a rock slide.

As the light from electric torches flooded the newly discovered chamber, the scientists found well-preserved footprints of several adult Neanderthals. Before this, little had been known of the physical appearance of these ancient troglodytes; now the impressions of their feet showed that they walked completely erect in the same manner modern people do. In fact their footprints were identical with those that a human of today might leave on a muddy path. The Neanderthal was truly in the family of Homo.

Nearby was the stump of a torch and, on the wall, a sooty, unmistakable human handprint. Radiometric dating by carbon 14, with an outer limit of 60,000 years, showed that this handprint was too old to date by this method. Scientists believe that the hand and footprints are at least 70,000 years old.

Footprints of humans from farther back in time become, not surprisingly, more and more scarce. To find any prints preserved from earlier cultures becomes increasingly rare, but such finds do happen. Near present-day Demirkoput, Turkey, an ancient hunter paused to rest from his perennial search for food just as a nearby volcano erupted violently, spewing forth fire and ash. The terrified man ran at breakneck speed for the nearby Gediz River. Sprinting over the soft volcanic ash, he left footprints, several of which were preserved. The eruption was probably of short duration, as the ash deposit is not thick. The hot sun baked the volcanic clay that bore impressions of his feet, and they were preserved as solid rock. Recently they were exposed through erosion.

The man must have run in panic, as the prints are almost a yard apart, and the toes were impressed quite deeply. We will never know if he made it to the river, because the prints that survived are over 250,000 years old. He was a specimen of *Homo erectus*.

The most spectacular story yet to unfold in our prehistory happened eons ago in East Africa at the onset of a rainy season. In that distant time the landscape stretched, much as it does now, into a series of savannas punctuated by wind-sculptured acacia trees. Not far to the east of the site lay the somewhat restless volcano now called Sadiman. It occasionally spewed ash over the flat expanse known now as Laetoli. Laetoli lies on the southern edge of the Serengeti Plain, the site of some of the most spectacular game migrations in the world. Its landscape 3.6 million years ago was surprisingly similar to today's typical East African savanna.

Animal life was quite abundant, and none of the numerous footprints suggest panic. Their wanderings were typically random in their routine quest for food, and they remained unshaken by the frequent eruptions of Sadiman. They had doubtless become accustomed to the rumblings and spewings of the neighborhood volcano. Several times it blanketed the plain with a thin layer of ash. Tentative showers, the precursors of the heavy seasonal rains yet to come, moistened the ash. Each layer hardened and preserved in remarkable detail the footprints left by the ancient fauna, none of which ever seemed to be in a hurry. These ancient deposits containing the footprints are geologically the oldest at Laetoli; they have captured a frozen moment of time from the very remote past—a pageant unique in prehistory.

The gray petrified ash beds hold the footprints of extinct elephants, hyenas, hares, large running birds, and, most important, a series of hominid footprints remarkably similar to those of modern man. The makers, among the ancestors of today's humans, laid these prints down an incredible 3.6 million years ago.

These exciting prints were discovered almost by accident. In 1976, after a hard day's work, two visiting scientists were returning to camp. Serious work was over briefly at the Leakey dig at Laetoli in northern Tanzania. The scientists, lacking play equipment, relaxed by throwing elephant dung at each other. As one scientist ducked to avoid a missile, his face came close to the ground surface of volcanic ash and he recognized a set of fossil footprints. That ended the horseplay and began a serious investigation of the area. By 1978 the prints were established as the greatest anthropological find to date. They were indisputable trackways of three hominids who lived here during the Pliocene epoch.

The volcano Sadiman periodically spewed forth great clouds of steam and ash. This was a particular kind of ash that contained carbonatite, a substance that dries into cement-hard layers when slightly wet. This was crucial in preserving these ancient hominid footprints. The event must have occurred when the dry season was giving way to the wet. As the excavations continued, impressions left by raindrops became common. The laminated ash clearly came from Sadiman in a series of short episodes, each represented by a single discrete layer.

Sadiman's periodic eruptions over about two weeks apparently weren't fierce or intimidating, for the antelope, hare, giraffe, fowl, and other residents of the ancient Laetoli community seemed to go about their business as usual. From time to time the landscape must have resembled a gray beach with scores of footprints impressed in the newly deposited ash. The ash, with its record of prints, was set and preserved into a rock-hard layer of volcanic cement called tuff. This would be followed by another eruption from Sadiman. More ash would fall. Thus these layers of hardened ash built up to a thickness of over six inches.

Late in the first week three hominids walked over the most recent ash fall and left their prints encased in the soft surface. It had rained just before their walk, because rain imprints are on the adjacent ash surface but not on their footprints. The sun was probably shining after the shower had passed. Shortly afterward Sadiman erupted again, and the prints were buried. The process was repeated several times before the rainy season began. By this time the print-bearing volcanic tuff was hardened cement; otherwise all would have washed away.

The footprints the three hominids left in the ashes definitely demonstrate that the earliest human ancestors walked fully upright with a bipedal (two-footed), free-striding gait. The tracks, nearly 60 footprints in all, run from south to north for about 80 feet before ending in an erosional gully. Two of the tracks run side by side. Judging from the size of the prints, the larger individual was a man who stood almost five feet tall. The smaller adult was a woman a little over three feet tall, and a third individual, the smallest, left its prints superimposed on those of the man. These prints seem to be those of a small child playing "follow the leader," just as chimpanzee young are known to do. The man must have walked slightly ahead of the woman. Their prints are so close together that side by side they would have constantly jostled and unbalanced each other.

One cannot help but reflect on this first nuclear family that took a walk in the sun nearly 3.6 million years ago. They could have been the ancestors of the family of Native Americans who walked in the sun along the Mojave River in California just over 5,000 years ago!

White—Not Always Right

Of all the disasters that can befall a traveler in regions of snow and ice, the main adversaries would seem to be cold, high wind, and avalanches. But there is a silent, apparently benign menace of a winter wonderland: the color white.

One result of overexposure to a peaceful expanse of dazzling, feathery snow or enchanting castles of ice is snow blindness. Ultraviolet rays reflected from snow or ice, or even white sand, can cause a temporary loss of sight. Normally the human eye can absorb this light, but under conditions of extreme exposure the eye's protective mechanism overloads and an abnormal intolerance of light develops. Victims have described the landscape as appearing pink and then red, and they feel an intense pain as though a handful of sand had been thrown in the eyes. Snow blindness does disappear after rest indoors, but it can be easily prevented by wearing dark glasses.

At times the polar regions are converted into a realm of fantasy where the senses become completely befuddled by the strange actions of cold air and drastic extremes in temperature. Light rays may bounce between the ice and low clouds, creating an opalescence called "whiteout." Every landmark is literally wiped out. To separate earth and sky and to determine where or if there is a horizon is impossible. As one veteran of whiteout described the experience, "It's like wandering around inside a ping-pong ball." This optical illusion can occur wherever snow covers large, flat areas. Although conceivably a worldwide phenomenon, whiteout is most common in the Antarctic and Arctic, areas where snow and ice are perennial. In scientific expeditions into the Antarctic, whiteouts are a much-feared weather phenomenon.

Although whiteouts can be caused by fog, fine precipitation, or blowing snow, they occur most often in clear, calm air under an unbroken layer of heavy, low-lying clouds that stretch from horizon to horizon. The uniform overcast diffuses the daylight and causes it to reflect between cloud and snow, thus obliterating all distinction between earth and sky. The horizon disappears, and nothing but white meets the eye. In the unbroken whiteness, judging depth or distance becomes impossible: what appears to be an oil drum 100 yards away may turn out to be a can of beans at arm's length—or vice versa. All shadows and surface details disappear in the unrelieved whiteness; walking is quite hazardous, since small holes and large crevasses are indistinguishable from the surrounding snow. A single step can result in a crippling injury or a fatal fall.

Vertigo, characterized by dizziness, disorientation, and disturbed balance, is also common during whiteout, so any sudden movement can become quite dangerous. With no horizon to serve as a visual reference, skiers often fall down. Some are so confused that they just lie where they land (actually a good idea). Drivers of tractors and snowmobiles often topple from their vehicles, and pilots, with no perception of depth, fly their planes headlong into the ground.

In 1958 at Ellsworth Base, Antarctica, a helicopter pilot caught in a sudden whiteout lost his bearings. Unable to distinguish just where the ground was, he tried to land the chopper at a 90-degree angle. The first thing that touched the ground was the nose of the ship, and it exploded. A rescue pilot almost joined the victim when his helicopter, which he presumed was flying level, bumped the ground at a steep angle. The pilot escaped from his demolished aircraft, but no one was able to rescue him until the whiteout had run its course. At present a standard practice for an airborne helicopter during a whiteout is to throw out some dark object on a line to help the pilot determine whether the craft is flying 20 feet above the ground surface or 20 inches.

Antarctic veterans have long concluded that the only sensible plan for whiteout survival while traveling on the ground is just to sit down and wait it out. Sooner or later the light will return to normal, and the details of the landscape will once again become distinguishable. The horizon will reappear to separate the sky from the earth, and travel can be resumed safely.

This Evolving Atmosphere

The astronauts who first stood on the moon were amazed at the incredible number of stars they could see in the black sky above. One can scarcely conceive the clarity of a night sky unimpeded by the layers of gases that make up the atmosphere encircling our earth. But there was a time when the sky as seen from Earth was as clear as that now observed from the moon. In the very primal stages of our planet's evolution, about 4.5 billion years ago, the earth did not have an atmosphere, and the stars shone very brightly against an extremely black night sky.

Scientists believe the first atmosphere was produced about four billion years ago. This primal atmosphere was released through active volcanism in the landscape of this primeval world. The original gases included water vapor, carbon dioxide, chlorine and sulfur compounds, and methane, nitrogen, and ammonia. Additional nitrogen may have been formed by the breakdown of ammonia (NH_3) under the influence of light.

Subsequently the water vapor condensed to form the oceans; the carbon dioxide reacted chemically with materials in the earth's crust and still remains locked up in many of the oldest sedimentary rocks. Of the original atmosphere, only nitrogen remains in high concentration (78 percent). The atmosphere has consistently changed with time, especially since humanity arrived on this planet and assumed control.

The dissociation of water molecules through discharges of lightning produced free oxygen in the primeval atmosphere some 3.5 billion years ago. By three billion years ago oxygen was becoming relatively abundant. Rocks and fossils of life forms of that early era indicate this. But the atmosphere was still not breathable by human standards. By the age of the coal-making forests, 360 million years ago, it reached a level of concentration somewhat similar to the present. During this, the Carboniferous Period, much of the land surface of the earth was an endless array of immense jungles and marshes. The intense activity of photosynthesis constantly released free oxygen into the atmosphere. As a result, the oxygen level by the time of the dinosaur, 100 to 200 million years later, was much higher than at present. Scientists determined this recently through the study of amber that had been in existence during the Cretaceous Period, 180–65 million years before the present (MYBP), the last stage of the dinosaur era.

Amber is the fossilized resin or sap secreted through the bark of prehistoric pine trees, similar to the fluid that exudes from conifers of today. The soft, sticky nature of the original secretions is indicated by the abundance of perfectly preserved insects, spiders, and even small lizards trapped in the semifluid sap. In time the resin hardened to become amber, which has been used as a gemstone since prehistoric man. Only recently has amber been prized by scientists for the record of ancient life that it can provide. Along with creatures, the resin preserved, most significantly, bubbles of air that would represent the atmosphere of 80 MYBP. This was the air that the dinosaurs breathed.

In 1987 two scientists obtained samples of amber that were 80 million years old. The air bubbles in the amber were a part of the atmosphere from the earth's Cretaceous Period. The scientists crushed pieces of amber in a vacuum and analyzed the air bubbles with a highly sensitive machine that can detect the concentration of various gases in extremely tiny samples. Their results were quite illuminating, for the amber proved to be inert and didn't contaminate the samples of air. As expected, the atmosphere for the dinosaurs was far different from that of today. In the preliminary analysis the oxygen level appeared to have been about 30 percent of the atmosphere, compared with 21 percent today.

A number of scientists now believe that the much higher concentration of oxygen is related to the giantism that characterizes many species of Cretaceous dinosaurs. Some stood as high as five stories, and the estimated weight of some herbivores was well over 100 tons. The abundance and variety of dinosaurs were also impressive. Their renaissance during the late Cretaceous seems to have followed the new, varied, and luxuriant plant life that could maintain a sizable population of large, well-fed herbivorous dinosaurs. This abundance could in turn support a plentiful array of giant carnivores, and all could enjoy the golden age of plentiful oxygen.

Amber of a more recent age, about 40 million years old, was obtained from the Baltic Sea area. Analysis of its air bubbles shows the oxygen level to be about that of today. Analyses of air bubbles 25 million years old show a similar oxygen content. The distinctly lower level following the time of the dinosaur requires an explanation, or at least a plausible conjecture.

Scientists generally agree that the late-Cretaceous dinosaurs were terminated abruptly by the collision of a gigantic celestial body from outer space with the earth, about 65 MYBP. Wildfires raged everywhere; the earth was truly scorched. When the fires died down, the forests of the earth were gone. The release of oxygen through photosynthesis being at a much lower scale, the oxygen content of the atmosphere dwindled to the present 21 percent.

For a study of the earth's atmosphere during more recent times, scientists have examined drill cores obtained from ice sheets of Greenland and Antarctica. The cores are significant scientific treasures, for they preserve continuous layered records of annual snowfalls going back about 125,000 years. The air bubbles trapped in the ice provide the same information as those encased in the older amber. The earliest of the ice cores were trapped during an ice age, and chemical analysis of the trapped air samples showed a somewhat lower content of carbon dioxide. This is significant because carbon dioxide is essential in retaining the warmth of the sun. The analysis also showed that the atmosphere contained a heavy burden of sunlight-screening dust, much of it volcanic. These two factors doubtless contributed to the ice age but were not its primary causes.

Our atmosphere today is becoming contaminated at an alarming rate. Recently scientists studied ice cores taken from a Yukon glacier about a thousand years old. They contained many trapped air bubbles. The research team found that the carbon dioxide level in today's atmosphere is 27 percent higher than it was before 1850, the beginning of the Industrial Revolution. And over one-third of this alarming increase has been within the last 25 years. Global warming, here we come!

They Also Serve

Few species in the animal world can boast of contributing a name, as well as a description, to one of the seven deadly sins. For the three-toed sloth, however, being slow (which is what *sloth* means) in all its movements and bodily functions is a virtue. Slowness made the sloth a valuable resident in the tree canopy of its Central and South American jungle home.

Mammals that have taken to the trees are usually characterized by agility, quick responses, and manipulative hands, feet, and tails—for example, the flying squirrel, all monkeys, the kinkajou, and even the opossum. The three-toed sloth made a radically different adaptation. With hooklike claws it hangs upside down from the topmost branches of the abundant crecopia tree or any of 90 other species of tree and vines. It gets all the food it needs from the foliage, flowers, and fruit of its tree, and sufficient water is available from moisture on the leaves. High above most predators, 30 to 90 feet from the forest floor, it blends unobtrusively with its environment. Remaining motionless and soundless is its way of attracting no attention.

Every feature of the sloth has evolved to make its life simpler and less demanding. Truly it has perfected slothfulness to both a science and an art. In so doing, the sloth defied many rules of how a body should be assembled. Lighter in weight than most animals its size, the sloth has very little muscle, a small heart, an oversized digestive system, a mere stump of a tail, arms much longer than legs, and a neck with nine vertebrae rather than the seven common to most mammals (including the giraffe). Its long, gray-brown hair parts in the middle of its belly and sweeps toward the back to accommodate its upside-down existence.

These physical adaptations simplify the sedentary, leaf-eating, arboreal lifestyle of the three-toed sloth. As it hangs suspended from a limb, rainwater can run off, and its short tail doesn't dangle conspicuously. Its head can turn 270 degrees (three-fourths of a full circle) to eat leaves to the rear or to observe the world with minimal movement. Weighing about eight pounds, it can crawl farther up or out on a limb than any of its pursuers. It can secure itself by a single set of claws and can climb easily and reach far with its long arms.

If an enemy does spot a sloth, what can happen? Hanging upside down puts it out of the grasp of most predators. Jungle cats have no chance unless they happen upon the sloth on one if its rare trips to the forest floor. Snakes have a hard time getting a grip on a sloth suspended from a branch. A harpy eagle may fly down with claws poised to pounce on the sloth, but the living pendulum's long hair and thick skin make it almost impervious to attack. Most predators have discovered that a firmly anchored sloth simply cannot be removed from its perch. Nor is it worth waiting for a better opportunity; weary jaguars give up when they realize that a sloth can outwait anything.

Living is uncomplicated although monotonous for the sloth. With little to fear and no struggle or competition for nourishment, shelter, or protection, the sloth's senses have become dulled and blunted. For an animal that depends on being mistaken for a bundle of dried leaves, this is an advantage.

Although the sloth's

activities differ little awake or asleep (while you have been reading this a sloth has done absolutely nothing), 18 out of 24 hours are spent sleeping soundly. The sleeping sloth hangs by its meat-hook claws with no fear of falling. In fact it must consciously unlock its grip to get loose and, according to legend, will remain hanging in place even after death. The sloth does find a change of position necessary when it has eaten all the leaves within its reach. There is also the sloth's regular trip to the forest floor.

Like the rest of the sloth's life functions, digestion proceeds slowly. But the bacteria in its multichambered stomach work constantly to break down the cellulose of leaves into energy. About every eight days the sloth descends laboriously to the base of the tree to urinate and defecate. Hanging on to the trunk with its forelimbs it digs a shallow hole at the base of the tree with its stubby tail. Here it empties bladder and bowels and covers the waste with dead leaves.

Since the sloth wisely does not share its tree, the trip to the forest floor is its only opportunity for a social life. Of course this is a risky adventure for an animal that can only drag itself on its belly when on the ground. But near a pungent dung heap seems to be the only place that a sloth can easily find another sloth, so it is probably here that a pair of sloths meet. They return to the tree for safety more than solitude, and they mate while, what else, hanging from a branch.

About six months later a single infant is born. Sharp claws clamped solidly to its mother's fur, it views an upright world while hanging on to the belly of its upside-down mother. Compared with adults, the infant is a bundle of activity as it climbs around on its mother, sampling her food and lapping water from her coat. If baby and mother become separated, they both call with a frantic "ai!" until they locate each other. This is the sloth's only sound, and in Brazil "ai" is the sloth's name.

Besides being a common trysting place, the sloth dung heap is a perfect place for many insects to lay eggs. Because the sloth returns to the foot of the tree regularly, hatched insects find adequate living quarters in its hair. Researchers have found several species of moths, beetles, ticks, and other mites living on the fur of the sloth. On a single sloth they found 120 moths and 978 beetles, all waiting for their host's weekly trip to the dung heap so they might lay their eggs.

The sloth provides the insects with their most basic needs; the insects merely provide the sloth with additional camouflage. The sloth's original disguise is provided by individual grooved hairs that are usually infested with algae, which gives the sloth a green hue. It can do no more to become part of its environment.

Several researchers have adopted sloths as house pets and learned much more about their life than can be exhibited by a motionless "wasp nest" or "ball of dead leaves" in a fork of a tree 90 feet from the ground. The fact that a sloth in a tree did not react to a gun fired nearby seemed to indicate quite blunted hearing. But longer, closer observation showed that part of its defense is to avoid reacting to any potential threat.

One naturalist placed a plastic disc on a sloth's head while it was asleep. The next morning, about 10 hours later, the disc was still there. On another occasion he noticed the smell of burning fur and discovered the sloth sitting, half asleep, atop a floor lamp. Its rear was smoking, but it continued to sit over the electric bulb. The slow reaction time of a sloth is not surprising; an animal with so little muscle has little energy to spare. Moreover the sloth rarely attempts actively to defend itself and will just make do with status quo. Most wounds a sloth receives slowly repair themselves.

Other sloth observers have seen them swimming over a mile to cross Gatun Lake in Panama. They are excellent swimmers and make the trip regularly. They are at home in water, enjoy a shower, and can be hung on the clothesline to dry while they bask in the sun.

A wildlife researcher kept several sloths as family pets during a four-year study in Suriname. The family was able to witness mating and the birth and rearing of an infant. They found sloths to be extremely adaptable as long as basic needs were met, such as a high place to perch and acceptable food. Sloths learned to accept alternatives to their favorite leaves: hibiscus blossoms, banana peels, cut-up citrus fruits, peanut butter, rice, and a great variety of greens. Although they rarely paid attention to each other, the animals displayed affection and enjoyed attention from the family.

Zoo experiences for the three-toed sloth have been short. Even with a well-duplicated climate and plenty of tropical leaves, it succumbs because of a dietary deficiency. Study of the minerals missing from their food while in captivity has identified the essential ingredient: minerals found in Cecropia ants, which are, for sloths in their home territory, a nuisance that accompanies every meal.

Considering that the sloth is not exactly an overachiever, how does it justify the niche it occupies in the ecological scheme? Actually, the sloth makes very few demands on its environment. Sloths make their homes in a variety of trees, thereby avoiding excessive pressure on any one species. Moreover, by burying their excreta at the base of trees that provide their food, they return to the soil up to 50 percent of the nutritional value of the leaves they eat.

The sloth neither preys on nor competes with any other animal. Its excrement is an ideal site for insect eggs, and it transports the adult insects in its fur, where they can live conveniently and safely. Sometimes doing nothing is exactly what needs to be done. Truly, the greatest service the sloth can render is to hang in a tree and wait.

Dinosaur Eater

Somewhere in New Guinea in 1944, a U.S. GI stood on a rock about 10 feet above an unusually large saltwater crocodile. He and three buddies were clearing the river of these reptiles as a service to the natives who were hiding the soldiers from the enemy. Earlier that day a native woman had been dragged into the water by a large crocodile, possibly the very one at which the soldier's rifle was aimed. The GI fired several shots at the crocodile, and it merely slid into the river. He would never know if it took to the water to soothe its bruises or if its injuries were terminal and it would never again dine on native women who washed their clothing and their children in the stream.

The saltwater crocodile, *Crocodylus porosus*, is the largest of all living crocodiles, reaching a length of about 25 feet. Although it lives mainly in the brackish waters of rivers and estuaries, it is not intimidated by the open sea. Individuals have been seen several hundred miles from land. This ability to swim considerable distances allows the crocodile to move from one island to another, so it has been able to colonize a large number of islands of the Sunda Strait in Indonesia and the western Pacific. It is also found on the coasts of continental Asia.

C. porosus has the reputation of being the most savage of human-eating crocodiles. It exceeds by far the record of the infamous Nile crocodile. It may, however, merely have a more advantageous position: near islands inhabited by people for whom surrounding waters are essential to their livelihood. A harrowing incident that occurred in a mangrove swamp on February 19–20, 1945, attests to the formidable savagery of *C. porosus*. Bruce Wright, a world-acclaimed naturalist on the island at the time, provided the most candid and reliable report.

A British combat unit had trapped about 1,000 Japanese soldiers on Ramree Island near Burma. Their only escape was through the swamp. The crocodiles, alerted by the din of warfare and the smell of blood, gathered among the mangroves, lying with their eyes above water, watchfully alert for their next meal. With the ebb of the tide, the crocodiles moved in on the dead, wounded, and uninjured men who had become mired in the mud. Their screams continued through the night, and by morning only 20 of the original thousand stumbled out of the swamp to surrender to the British. This was doubtless the most deliberate and wholesale attack on humans by large animals ever recorded.

The ferocious saltwater crocodile has the patience and ability to become almost invisible to an unsuspecting victim. Lying in or near the shores of rivers and lakes with only its eyes above the water, it looks just like a floating log. A thirsty animal who comes to the river to drink will be seized in scarcely an eyeblink. Given the uncanny strength of the reptile, the victim rarely has a chance. It is quickly pulled into the water and held beneath the surface; if the victim had a choice, it would prefer to succumb to drowning before the crocodile begins to snack on the fresh catch. Recently in Australia several people witnessed a 15-foot crocodile grab a very large horse by the hind leg and drag it into the river with very little effort.

Even if saltwater crocodiles had not killed and dined on many humans in Australia and the East Indies, they would be an object of dread because of both their menacing tooth arrangement and their enormous size. But today's saltwater crocodile is a miniature model of its dinosaur-eating ancestor that inhabited rivers 100 million years ago in what is now North America.

During the Cretaceous Period, the time of the dinosaur, the waters of many rivers were host to an ancestor of *C. porosus*. This ancient crocodile, *Phobosuchus*, lived a lifestyle similar to that of modern crocodiles. Hidden in slow-moving waters, with only its eyes exposed to view, it would wait for some thirsty dinosaur to come to the river to drink. It likely struck with the same incredible speed its modern descendant exhibits. Grabbing the dinosaur by a leg, the Cretaceous crocodile would drag it into the water with no more effort than was expended by the crocodile in Australia as it subdued a horse 100 million years later. The prey of *Phobosuchus* were enormous; probably even dinosaurs similar to the celebrated *Tyrannosaurs rex* fell victim to this denizen of the ancient rivers and lakes. After all, *Phobosuchus* sported a 6- to 7-foot skull and was over 50 feet long!

Who Were the Real Barbarians?

Most people associate pyramids with Egypt and northern Africa, where almost 300 have been excavated. But pyramids were widely distributed among ancient peoples. The New World has larger pyramids, and more of them, than Egypt ever had; they number in the thousands in Mexico and Central America.

During the 2,500 years before the arrival of the Spanish conquistadors, the Mayans developed a civilization that dominated Central America. When they abandoned their cities, for complex reasons still being unraveled, they left behind some of the most impressive and mysterious architecture. It is mysterious because their hieroglyphics are far from understood, and no Rosetta stone has been discovered that would help to decode them. Since the Mayan cities were abandoned, the dense foliage of the rain forest has buried more pyramids than existed in all of ancient Egypt; only an estimated 15 percent of the Mayan pyramids have been uncovered.

The Maya appeared in the land 3,000 years ago, building a culture that flowered while Europe languished in the Dark Ages and that survived six times as long as the Roman Empire. They were at the zenith of their development around A.D. 300. Their cities were far larger and more elaborate than anything in Europe at the time. Their complex and sophisticated culture lived by a calendar equal to that used today by modern western civilization. They independently developed the concept of zero in mathematics, predicted eclipses of sun and moon, and traced the path of Venus with an error of only 14 seconds a year. They reached the highest achievement throughout Central America in engineering, astronomy, stone carving, and mathematics, and they were the only group in the hemisphere to have a written language. Priceless records of history and cultures were kept in libraries.

The coming of the Spanish was their final undoing. A Mayan prophet had predicted that men with beards would come from the east, and so they did, beginning in 1517. The conquest began with ruthless slaughter by Pedro de Alvarado, to whom Cortez had assigned the conquest of Guatemala. Although Alvarado had been directed to bring the people into the faith without war, he pursued them relentlessly with the power of swords and guns rather than with gentle persuasion. Another conquistador, Francisco de Montejo, received permission from King Charles to conquer Yucatan at his own expense. It proved to be a very expensive effort and never was completely successful. The first two attempts, in 1527 and 1530, failed. The third and relatively successful invasion began at Can Pech, now Campeche, the oldest surviving European settlement in the Yucatan. The official surrender took place at Ichcansico, now Mérida, in 1542.

The all-conquering Spaniards looked on the Mayans as barbarians, although they didn't hesitate to confiscate Mayan gold and precious jewels. The practice of human sacrifice was an appalling part of the Mayan culture, and the "civilized" Spanish conquerors were horrified. In fact the Spanish were so taken back by the Mayan practices that

Bishop Diego de Landa ordered all the records of "barbarism" burned. His orders were carried out immediately. Within hours almost all written records of a great civilization were gone.

Only three codices (manuscript volumes) of Mayan writings somehow escaped the holocaust. Even more ironic is that about 90 percent of the soldiers burning the records of Mayan history and culture were themselves illiterate and could neither read nor write. But they were skilled in conquest, slaughter, and destruction. Just who were the real barbarians?

The Moth and the Candle

"Thus hath the candle singed the moth."
Shakespeare, *The Merchant of Venice*

In a scene that set the stage for the 1982 motion picture *Quest for Fire*, a Neanderthal man is sitting near a campfire in front of a cave. Inside the cave the rest of the tribe is asleep. His job is to keep the fire burning throughout the night, for if it were to die out they would be without their most prized possession: fire, their most effective weapon and their source of light and warmth. Although they understand its uses and how to keep a flame alive, they do not know how to start a new fire. Their only solution to being without fire is to steal it from another tribe, a most dangerous quest. As the man sits watching the fire, he is intrigued by a moth flying in rapid circles around the flames. With a fast arm maneuver, he catches and promptly eats the moth.

Had this Stone Age man waited, he would have seen the moth narrow its circles around the fire until the flames consumed it. The moth would seem to have deliberately flown into the fire. This action of the moth around a flame has been observed by people of all times, from early beings up to the present. The popular opinion is that the moth becomes so hypnotized by the dancing flames that it irresistibly flies into the fire. The tendency of moths to fly into a fire has been studied intensively by leading entomologists, who use a candle as their basic research tool—hence the common pairing of "the moth and the candle."

The moth's behavior is neither suicidal nor supernatural. Actually it is a victim of a situation that rarely arose until after humans appeared on the earth. By this time the insect had evolved a tropism (an innate tendency) that would direct it to respond to an external stimulus such

as light. During the millions of years before humans started lighting campfires, the moth was rarely at risk from a source of light.

A moth in night flight will beat its wings faster when light, such as from a campfire or a candle, suddenly falls on its eyes. If the light of the campfire is off to one side, it falls more strongly on one eye than the other, and the wings on the side nearer to the bright light will tend to beat faster than those on the other side. Thus the moth flies in circles around the light source. The closer it gets to the flame the stronger the light becomes, and the faster the wings beat. The pattern of the moth's flight causes it to fly in ever-smaller circles closer and closer to the source. Inevitably it reaches the flame, and ZAP!

The moth appears to have deliberately committed suicide. But actually it is the victim of one of the mistakes that nature made in developing this tropism, for even nature could not foresee every eventuality.

Trap–Door Spider Versus the Wasp

Several species of trap-door spiders protect themselves against predatory wasps by digging long, silk-lined chambers in the ground for their homes. They cover the holes with snug-fitting doors ingeniously camouflaged and carefully concealed under plants.

The female wasp who does locate the trapdoor to the underground nest knows better than to venture into the hole. She is instinctively aware that the spider would have the advantage in its dark home and would instantly kill her. Instead the wasp waits like a stalking cat in front of the hole. When the spider feels its way out, the wasp seizes it by the forelegs, swings it out in an arc, and stings it.

Many trap-door spiders provide a second exit from which to observe what the wasp is doing at the first hole. The clever wasp, however, is often prepared for this. She will insert her hind part a short way into the first hole and then hurry to the second, where she catches the emerging spider. If this fails the first time, the wasp repeats the maneuver until she succeeds.

The successful wasp injects her prey with a venom that paralyzes but does not kill. Then she drags the victim to a secluded spot and lays eggs on its still-living body. When the eggs hatch, the newborn will have fresh spider meat to feed on.

Dust to Dust

During a late afternoon in June 1962 a geologist mapping an area called the 111 Ranch Beds near Safford, Arizona, took a break. As he sat admiring the desert scenery, he observed a high wall of dust—the front of a violent sandstorm about a mile to the south. In a few seconds he jumped to his feet and yelled, "Holy smoke, it's coming this way!"

He reached camp, which consisted of a field vehicle, just as the storm hit with great impact. His partner was not present, and he worried enough to tie a hundred-foot rope around his waist with the other end attached to the door handle. Wearing wind goggles he crawled on his knees into the storm trying to find his friend. Reaching the end of the rope he started back to the truck, taking a slightly different path, and ran smack into his partner, who had tied a rope around himself and was looking for partner number one. With visibility almost at zero, they followed the ropes back to their vehicle and sat out the storm, which lasted for at least an hour. Both men, badly sandblasted in the face and arms, had to return to Tucson for medical care.

As climatic historian Paul Sears said in *Deserts on the March*, "Dust is nothing new. . . . Like the circle, it is a symbol of eternal time." For millions of years atmospheric winds have picked up dust and carried it elsewhere. Throughout the world are vast deposits of fine sediment often hundreds of feet thick. Geologists refer to these thick deposits of dust as *loess*. In other places sandstone layers with sloping bedding planes attest to wide expanses of desert throughout hundreds of million of years of time, even in places where today the climates are humid and tropical.

Extremely violent sandstorms have always occurred in the deserts of the world. In 1923 a U.S. expedition team was busily excavating a dinosaur nest in the Gobi Desert. To the scientists' amazement they uncovered the skeleton of a four-foot-long toothless dinosaur lying about three inches above a clutch of eggs. Later named *Oviraptor* (egg seizer) the animal probably lived by feeding on dinosaur eggs. This unlucky specimen was in the act of digging up the nest when it was overcome by a violent sandstorm and buried alive on top of the very eggs it had come to steal. This prehistoric drama occurred over 100 million years ago.

Some of the earliest journals of civilization call attention to dust clouds—if only for purposes of reverence, even religion. On the Thracian plain of Greece, Dicaeus the Athenian and Demaratus the Lacedaemonian once chanced to be standing where a great cloud of dust advanced toward them, not unlike the dust cloud observed by the Arizona geologist thousands of years later. To these two ancient Greeks it was like a cloud of dust raised by 30,000 men marching, and they wondered who the men might be. Suddenly they heard the sound of voices, very likely the howling of the wind—but to them it was voices singing the mystic hymn of Bacchus. As they watched, the dust became a cloud and rose into the air to sail away to Salamis, an almost sure prediction of ill for their enemies and good for themselves. No doubt the enemy had a reverse opinion.

Both Homer and Virgil mention dust storms. Early legends tell of shifting sands burying entire caravans and marching columns. Violent sandstorms and dust storms are not uncommon in the great Sahara Desert, but people born into this environment have learned to cope with them as normal events. Sahara storms can be much more severe than those in Arizona and will last much longer, sometimes for days. Some of the most violent storms have been known to drive people mad or, at times, bury them alive as it did the dinosaur 100 million years earlier.

Sand and dust are almost always on the move; they are picked up, blown a short distance, and dropped to earth again. But at times great amounts of dust are carried thousands of miles to settle in foreign lands. Some Sahara dust storms have become exceptionally widespread: in 1901 a mass of dust estimated to weigh nearly four million tons was spread in a thin mantle across North Africa and Europe. An estimated 10 million tons fell on England in two days in 1903, and it is not uncommon for ships at sea to be brushed with dust from the Sahara.

Dust on the move can produce some rather unusual effects. Red rains and red snows, comparatively frequent in some places, are the result of reddish Sahara dust encountering humid air and condensing to douse broad areas with reddish sediment. Some of the millions of tons of Sahara dust sailing along at high elevations to moist climates is precipitated widely, as occurred in Germany in 1959. Over 15,600 square miles were blanketed with deep reddish snow. Dust seems to prefer that area; in 1980, ski resorts advertised the thrill of skiing on red snow. Business zoomed!

The wind current that flows off the Sahara Desert is known as a sirocco. It is responsible for delivering the unusual precipitation as it carries million of tons of reddish-colored sand over Europe. However, such events can have negative results. In A.D. 582 a deep "shower of blood" fell on Paris. The red rain caused great fright; many Parisians tore at their "bloodstained" clothing while vigorously repenting their sins. The local churches were filled to capacity that day.

Windstorms can have a comical side. In 1901 an engineer examined a prospective placer diamond property in what was then German East Africa. He found the sand rich in very fine diamonds and began negotiating for a deal with the owner. When a fierce windstorm suddenly descended on the area, they retreated into the owner's shack. Both watched with mouths agape as the wind picked up the sand, stripping the area down to bedrock, and carried sand, dust, and diamonds to other properties farther down the valley. The deal was off.

CHAPTER SEVEN

Herds of Dinosaurs and Bison

In 1978 western Montana was the site of a remarkable discovery: the fossilized remains of 15 baby *Maiasaura* ("good mother lizards"). The presence of numerous eggshell fragments and of many very young dinosaurs indicated a nest. The young dinosaurs had well-worn teeth and were too large—three feet long—to have been newly hatched. The implication is, therefore, that the young stayed together for some time because they were being cared for by their parents. Food would have been brought to the nest, for if the young left to search for food, they would not likely return to a nest abandoned by parents.

This picture of dinosaurs is not surprising. Crocodiles, distant relatives of the dinosaurs, show considerable care of their young. The parents move hatchlings to nursery pools and guard them against predators until they are large enough to defend themselves.

Further excavation and exploration of the Montana nest sites uncovered numerous nests at different levels, evidence that the *Maiasaura* returned year after year to rebuild their nests and raise their young. Each nest lay about 23 feet from the others, the length of an adult. The nesting colonies must have been enormous because the herds migrating to this area were, according to recent discoveries, also enormous. Scientists have uncovered the remains of an entire herd in the vicinity of the nesting site; all died simultaneously 80 million years ago.

Under the direction of John Horner of the Museum of the Rockies in Boseman, Montana, seasonal exploration and excavation were continuous during the 1980s. Researchers uncovered a massive bed of maiasaur bones, with an estimated concentration of up to 30 million fossil fragments covering an area 1.25 miles wide and .25 mile long. Without exception all of the bones were of *Maiasaura*. These were the remains of a single herd, conservatively estimated at no fewer than 10,000 dinosaurs.

The layer of volcanic ash that rested just above the layer of bones tells us what happened. The eruption of Mount St. Helens in 1980 produced a widespread, devastating ash fall; such volcanoes were run-of-the-mill in the Rockies of the *Maiasaura*'s time. The entire herd fell victim to the gases, smoke, and ash of a huge eruption. Since researchers found no evidence of scavengers or predators gnawing on the bones, the eruption doubtless also killed everything else around. Imagine a feeding carpet of dinosaurs staring in confused disbelief as an enormous choking black cloud of volcanic ash swept over them. In a few seconds there remained only a huge killing field strewn with the corpses of 10,000 dinosaurs under a soft cover of volcanic ash. Soon the stench of rotting corpses would have been overpowering. Until the ever-present insects cleaned the remains, it must have been one heck of a mess.

Perhaps surprisingly such a great herd of maiasaurs survived on just sedges and berry bushes, common vegetation of that time. By comparison the bison, which numbered at least 60 million in the early 19th century, ate nothing but grass. The demise of the maiasaur herd was small stuff compared with the fate of a larger, more recent herd of bison. They also lived in the state of Montana but 80 million years later.

The American bison, incorrectly called buffalo, measures 60 to 70 inches at the withers and weighs almost a ton. It is a powerful beast, and the slightest movement of its shoulder muscles reveals a fantastic potential energy that defies comparison to that of any other living creature. Bison enclosed behind huge tree-trunk fences may remain docile, almost immobile, prisoners for months. Then suddenly, with no apparent warning, they will smash the barrier as if it were made of matchsticks.

During the mid-1800s the bison herds were so large that trains were known to wait for days as an entire herd crossed the railroad tracks. Their numbers were so vast that nobody thought or believed they could ever become extinct. Perhaps even nature thought so.

In February 1858, Montana's unusually harsh winter seemed to reach a climax. A blinding blizzard had been raging for days. The migrating herd of bison knew instinctively that they must keep moving, and move they did. None dared lie down to rest. The vanguard of the herd perceived only a whirling, swirling mass of snow. The howling of the wind was continuous and deafening, and visibility varied from a few feet to a few yards. With the entire landscape blotted out and blended into the white opaque air, the bison headed blindly against the storm and swirling wind. They stayed very close together, and each individual could see only the hind legs and rumps of bison immediately in front of it.

The frontline bison received no warning, no variation in color of the snow or furious movement of the snow-flecked air to alert them to any change in terrain. Even had they been aware of the precipice, however, those in front could not halt because of the formidable pressure of thousands of animal engines behind. Like a carpet shoved off a ledge, the entire herd streamed over the precipice.

Nothing could have prevented this catastrophe. The wind howled so furiously that the animals couldn't possibly hear the noise of those before them tumbling one over the other from a height of more than 150 feet. Without exception, all 100,000 of them walked off the cliff to their deaths. The vast assemblage of their carcasses spread over hundreds of acres. It was a true killing field—and eventually a boneyard awesome even to seasoned men of the mountains.

In the days when bison herds roamed the west, the loss of 100,000 bison at one strike was of little significance. They had no real enemies except wolves, Native Americans, and the elements. One herd might have disappeared, but a thousand other herds survived. And so the bison lived on through all seasons, overcoming all difficulties, surviving for centuries. It was as though the American bison were immortal.

Then one fine spring day the white hunter arrived.

Requiem for a Hero

The gorilla, villain of so many movies, books, and TV shows, is in reality a very gentle creature. The antics it puts on when threatened—howling, jumping up and down, beating on its chest—are mainly bluff. When it charges, just don't run away; it will stop, turn around, and walk away. This is incredible, considering how powerful a full-grown male gorilla can actually be.

Only when the male gorilla feels his family is threatened will he become the aggressor. Then he will exhibit great courage. Recently a group of U.S. scientists were studying a 14-member family of mountain gorillas. The animals were suddenly attacked by poachers unaware of the presence of the scientists. One huge male gorilla, as the largest in the group, placed himself between his fleeing family and the attackers, who just did not expect this type of behavior. The battle that followed was furious, but the gorilla managed to stave off the attackers while the family escaped. During the heat of the fight he never once tried to escape. Unfortunately he took five spear wounds during the fight and did not survive.

The poachers spotted the Americans after the gorilla was killed. They promptly fled, leaving behind the carcass they had tried so hard to poach. The scientists, moved by such courage and devotion to family, carried the dead body back to their camp. Considering the gorilla weighed several hundred pounds this was not done without difficulty. They wanted to honor the fallen animal, and so they gave him a formal burial.

Does the Menu Bug You?

In many parts of the world, insects are an important component of everyday diet. Among early and primitive peoples, who bore no prejudice against creatures that crept, crawled, and buzzed around them, a handsome grub or beetle was a tasty, accessible snack food. The insect eaters were not aware that insects, egg and larva as well as adult, had the highest percentage of protein of any natural food. A cricket, for example, is 50 percent protein, compared with lean beef at less than 20 percent.

The bias against insects as food falls off sharply when people are starving. During their exodus from Egypt, the children of Israel wandered hungrily across the desert, praising God for manna from heaven that was spread daily across the face of the land. Manna turned out to be a secretion produced from the digestive juices of a scale insect combined with the sap of trees still common in the Sinai peninsula. Other biblical favorites were the locust, beetle, and grasshopper, which were among the acceptable foods identified in Leviticus. In a list where many beasts, fish, and fowl were designated as unclean and an abomination, these insect families were given a clean bill of health.

Since Assyrian times, when locusts were a delicacy served to monarchs, they have remained a significant part of the Arab cuisine. Sun-dried, roasted, grilled, boiled, prepared in couscous, fed to camels, they have remained an important food throughout the Muslim world. Where locusts are overplentiful, the choice is either allowing them to ravage crops or making them the crop. In 1875, when locusts were plaguing the United States, an entomologist delivered a scientific paper suggesting they be used as food. He prepared a supper of locust soup, locust cakes, and grilled locusts, with baked locusts and honey à la John the Baptist for dessert. His idea, along with his menu, proved to be quite acceptable to the scientists who shared the meal.

Among the native people of Australia, where all nature's creatures are deemed potentially edible, the women gather beetles, caterpillars, grubs, and honey ants and prepare them for the next meal. A more impromptu snack consists of such creatures as lice that are currently freeloading on their bodies. The Inuit, unable to keep their warm animal-skin garments free from lice, also eat everything they are able to collect. Head lice are a shared delicacy among some native people of the Amazon.

In many parts of Africa, more than 60 percent of dietary protein comes from insects. Some people in Ethiopia live almost exclusively on grasshoppers; others dine on plentiful giant mosquitoes. Ordinary intolerable, unswattable mosquitoes are considered too low in food value to be worth collecting. Moreover, a meal of plump mosquitoes that have just nourished themselves on human blood may seem slightly cannibalistic.

Termites are a favorite among many African people, and collectors have perfected the art of trickery. The termite hunters, having noticed that termites leave their nests in great numbers when it rains, surround a termite hill and imitate the sound of rain. As the hunters beat gently on pots, the termites stream out of their holes and are swept into tubs of water. Then they are ready to be grilled, crushed, shaped into cakes, rolled in banana leaves and poached, making a succulent and nutritious termite sausage. Hottentots eat termites boiled or raw, and other Africans roast and eat them like jelly beans.

Latin America is considered a world capital of entomophagy (insect eating), with creative variations on any number of insects, caterpillars, and other larvae. A bowl that appears to be steamed rice is more likely escamoles, the larvae of red ants. Children shake nuts out of trees and break them open to find a curled-up grub within that is sweeter than the nutmeats. At movie theaters roasted ants are eaten like popcorn. Crickets have crossed the Rio Grande into the U.S. Southwest. They are roasted until crunchy, seasoned with a salt, garlic, and cayenne butter sauce, and known as Crispy Cajun Crickets.

Recently Chinese scientists developed an extract from maggots of the housefly and are negotiating with food and drug firms to mass-produce the nutrition-rich product. From just 1,000 grams of maggots can be extracted 500 grams of pure protein and 200 grams of low-fat oil and amino acids. Certainly this could be a valuable dietary supplement for the people of the world. One might hope that truth-in-packaging laws will allow a euphemism for "maggot" and leave the potential user no clue as to contents. After all, the unlisted "ingredients" that inadvertently slip into a bottle of catsup or a package of chicken are no cause for alarm.

Many creeping things are used as food by one or another population, and squeamish, narrow-minded westerners are no longer startled by what other people of the world dare to eat. A recent banquet of the Explorers Club included hors d'oeuvres of fried mealworms, crickets, and wax worm fritters. The centennial banquet of the Entomological Society featured live honey ants, giant Thai water bugs, roasted crickets, and chocolate cricket torte. As emigration from Southeast Asia and Mexico increases, insect cuisine may become more acceptable in the

United States. Within the next decade, bugs may be displayed rather than hidden on grocery shelves.

From Out of a Cloud

All nine of the planets in our solar system follow elliptical orbits around the sun. Just as important, they all lie in approximately the same plane.

Because of this, many scientists conclude that the sun and its planets were formed simultaneously out of a cloud of cosmic gas and dust. Scientists are quite sure that if the planets did not come into existence thus, they would circle in different directions along their individual planes, like moths fluttering around a candle, each in its own orbit.

A Dream of One World

During the fourth century B.C., in the mountains of the Balkan Peninsula slightly beyond the reach of Greek civilization lay the semibarbaric kingdom of Macedonia. In the city of Pella, Macedon, in 356 B.C. a son was born to King Philip II and his wife Olympias, a princess from Epirus. His name was Alexander III, to be known to future generations as Alexander the Great.

Time, place, and birthright were favorable for Alexander. Immersed in his father's claim that he was descended from Heracles (Hercules) and with his mother's ancestry traced back to Achilles, Alexander felt predestined to become the most successful conqueror the world had ever known.

At the time of Alexander's birth a veneer of Greek civilization was masking the rough, uncultivated population of Macedon. Philip, educated in the Greek tradition, demanded the best for his son. He summoned Aristotle to the Macedonian court to instruct the young prince in Greek literature and lore, science, and logic. Alexander was an apt pupil, and as his entire character became immersed in the splendor of Greek genius and culture, he could easily imagine himself another Homeric hero destined to fulfill a prophecy.

As Alexander was approaching manhood, he learned from his father the art and science of warfare. From the peasant Macedonian population, Philip assembled a standing army of professional soldiers, added a cavalry unencumbered by chariots, and advanced as an invincible machine with all parts working in unison. He began his conquests in regions east and north, where he would meet the least resistance. Finally Philip was able to unite the Greek city-states into a federation of Greek allies, with himself as leader rather than conqueror.

Philip was preparing to embark on a mission into Asia Minor, where other Greek cities would be freed to join the federation, when he was stabbed and mortally wounded. Power passed into the hands of Alexander, a youth of 20 years. He continued his father's work with the assistance of remarkably capable and devoted leaders from the Macedonian court.

The Greek states became rebellious against Macedonian leadership and determined to overthrow the youthful ruler. Thebes, the first to revolt, was destroyed by Alexander. Only the house of the great poet Pindar was spared. All of Greece learned quickly to respect young Alexander's power.

Alexander moved directly into Asia with armies augmented by troops from the league of Greek states. He camped at the site of the Trojan War to absorb the aura of Achilles. At Gordium he listened to the legend revealing that whoever undid the Gordian knot would reign over the entire East, and he promptly cut the knot with his sword. He continued eastward to conquer Persia under King Darius III. He finally crossed into India, conquering until his victories encompassed the breadth of the civilized world, from the Adriatic Sea to beyond the Indus River.

According to legend, when Alexander realized that there were no more worlds to conquer he cried. This seems unlikely. When a "longing came over him," as his biographers observed, he would plan another conquest. Several semibarbaric lands to the west still had not met Alexander the Great, and they never did!

Alexander founded 18 new cities in his two million-square-mile empire. At least 10 were named after himself, one after his dog; one at the site of a river crossing where his horse Bucephalus was killed he named Bucephala. His interchange of science (he sent hundreds of natural history specimens to his old teacher Aristotle), commerce, arts, and architecture had a lasting impact on Greek as well as Oriental civilization.

Alexander recognized that to hold together a single world of many different nations the people would have to intermingle freely without being poisoned by racism. The unification of his vast empire could be best achieved by intermarriage between the conquering and the conquered people. He put his theory into practice by arranging mass marriages between his Greek and Macedonian troops and the people of Persia, Egypt, Ethiopia, Assyria, India, and every other nationality in the empire. His own wife, Roxane, considered the most beautiful maiden of all Asia, was a captive from eastern Persia. He planned to continue widespread intermarriage until all people would be one race, one nation, one melting pot.

Alexander's ambitious plan might have been successful except for an unforeseen bite from a mosquito. At age 32 he suddenly became ill and died within a few days. The symptoms resembled malaria, and this seems the most logical of possible explanations. Having no son to carry on his reign as king of the world, he was asked on his death bed to whom he would leave his kingdom. He is said to have whispered, "To the strongest." No successor was named, and his generals began to fight among themselves while carving up the empire. Alexander, and his empire, crumbled to dust that blended with the dust of those who died for his greatness. Only thus, in the powdered earth of mingled remains, did he achieve his dream of one world.

Shark Prank

Thousands of sharks are caught annually by fishing enthusiasts as well as by those who hunt sharks for a livelihood. Surprisingly, only a few dozen serious injuries have been reported in the last 25 years. The following mishap could have had an even more disastrous outcome than it did—no fault of the shark!

During July 1970 a sport fisherman off the Virginia coast, already disgruntled with the day's catch, hooked a small five-pound sand tiger (a shark) and dragged it aboard. This was quite a letdown for a man who had visions of bringing in a great white. Less than thrilled with his catch, he stuffed into the shark's mouth a small bomb of the type used to train soldiers, set the timer, and tossed the shark overboard. (Just how he happened to be carrying such a bomb seems irrelevant; no one bothered to ask.) The fisherman and his fellow sailors considered this the joke of the day, and all rushed to the railing to see the little shark blown to pieces.

As with many practical jokes, this one backfired. Instead of swimming away, the little shark doubled back and, with flawless timing, blew up squarely underneath the boat. The explosion blew a large hole in the bottom planks, and the boat promptly sank. The disgruntled owner, who faced a $5,000 salvage and repair bill, sued the fisherman for a considerable sum and was awarded the entire amount.

The story does end here. The bomber fisherman retired permanently from shark fishing.

Not So Cuckoo

For many centuries the cuckoo has been regarded in a variety of contradictory ways. It has been the herald of spring and a symbol of good fortune. As a voracious eater of insects, especially the larvae of many species harmful to crops, it has been considered a friend of the farmer. Its name represents the loud insistent call of the common cuckoo of Europe and Asia; the word for cuckoo in many languages is also a phonetic representation of its call.

The slang use of "cuckoo" to mean crazy or foolish seems appropriate for a noisy creature with such a reprehensible lifestyle, for the cuckoo is best known as a self-serving, insensitive, ruthless parasite. Not only does it not build its own nest, but it lays its eggs in the nests of other birds, dumping an egg of the nest builder to make way for its own. The nest builder becomes the foster parent charged with the task of brooding, feeding, and raising the squatter to maturity. Meanwhile the mother cuckoo has taken off for a trip of 2,000 miles or more to an ancestral wintering place.

Of the 127 species of cuckoo, fewer than half are wholly parasitic. These include the Old World species of Europe, Asia, Australia, and the Pacific Islands. Those of the Americas usually build nests (however shabbily), brood, and care for their young. An exception is the American striped cuckoo, a typical parasite whose eggs are deposited for adoption.

The striped cuckoo is best known for its melancholy, whistlelike call and because it doesn't know when to shut up. People subjected to its song for hours would doubtless like to quiet it permanently. However, the striped cuckoo has a very effective defense: a talent akin to ventriloquism. By varying the amplitude of its whistle while turning its head, it gives the illusion that it is moving about. In spring 1992 an ornithologist attempting to record the birds whistling to each other felt he was stationed near enough to the bird to tape the song, because the whistle seemed about 10 feet away. He was absolutely astonished to discover that the elusive but stationary bird was actually several hundred feet from him. The frustrated ornithologist was never able to get close enough to activate his recording instruments effectively.

The common cuckoo of Eurasia is typical of the legendary parasitic cuckoo. The female's most advanced parenting skill is her selection of a proper foster parent, thus ensuring that her offspring will have a good chance for survival. She keeps watch on a smaller bird, typically a perching bird, as it builds its nest. Immediately after the foster parent completes the nest, lays her eggs, and flies off in search of food or relaxation, the female cuckoo flies down. She lifts an egg out of the nest, swallows or drops it, and lays one of her own in its place—all within about eight seconds. Then she is off and away!

The common cuckoo, which winters in Africa or India, begins her migration immediately after she has deposited her last egg in a host parent's nest. She will never meet her offspring, but she can be confident that they are receiving the best possible care.

The parasitic egg is safest if it resembles the eggs of the host in color and markings. The natural parent has also been careful to select one of over 300 species of birds that nest high in trees and do not begin to brood their eggs until all are laid. Interestingly, each female cuckoo chooses a single host bird species for all the eggs she lays. Possibly when she was reared by her own foster parents she was imprinted with the type of nest, eggs, and bird that had been adequate to her needs, and she retained this impression for life.

The returning foster parent seems not to notice that one egg is larger than the others. Occasionally clutches of eggs containing a cuckoo egg will be abandoned by the foster mother, but usually she obeys instinct and accepts whatever is in the nest. The cuckoo egg hatches in $12\frac{1}{2}$ days, a half-day or more before its nest mates. This, along with its larger size, gives it a decided advantage over the natural offspring. The newborn cuckoo will first attempt to evict the other eggs or hatchlings; it maneuvers itself to the bottom of the nest, balances an egg in a depression between its wings, and hoists it overboard. Any that survive being tossed out will be ignored by the parent, for her only concern is the occupants of the nest.

The baby cuckoo is usually successful in removing all competition for food. Even if any eggs or babies remain after it cleans house, they will surely be no match for a larger, older, rapidly growing cuckoo. A widely gaping brightly colored mouth triggers the parent to drop food in, so nest mates will probably starve.

The foster parent is hard put to keep the young cuckoo fed, and it becomes more unwieldy as it grows. By about three weeks it is too large for the nest and is soon larger than the parent. A grotesque caricature of bird family life is a tiny host parent perched on the back of a large, noisy foster child, trying to push enough insects and grubs down its gaping beak. The well-chosen parent continues to nurture the offspring until it is fledged and ready to fly.

Most amazingly, the young cuckoo makes a completely unguided flight across 2,000 miles of unknown territory. Both the flight path and the destination could have been reckoned only by instinct. After that, even a chance meeting with its natural mother in their ancestral home should not be surprising.

Legendary Earthquakes

During the 17th century the Jamaican town of Port Royal was a hangout for pirates and all manner of criminals—definitely a haven for the underworld. Until June 7, 1692. On that day the city was struck by a devastating earthquake that destroyed more than two-thirds of Port Royal and killed over 2,000 of its inhabitants. The survivors saw this as a divine warning. Most major crime virtually ended forever.

The earthquake caused the crime-ridden northern port of the city to slide into the sea. Perhaps as a grim reminder of "sin-city days," intact buildings can still be seen under water!

Scientists now appear to know the exact time of day the earthquake occurred. In 1956, from the submerged ruins, research divers recovered a coral-crusted pocket watch made in 1686. Laboratory x-rays later revealed that the steel hands had stopped when the watch made contact with the water at 17 minutes before 12. The moment of the earthquake was thus recorded on the timepiece of one of the victims, assuming the watch was keeping the right time. What still puzzles the scientists is, was the watch recording A.M. or P.M.?

A similar, earlier incident was also the result of a tectonic event. In 1967 U.S. scientists investigated a recurrent Indian legend of sunken ruins on the floor of Lake Titicaca, Bolivia. According to the legend, an intense earthquake caused the lake waters to inundate some pre-Inca temples built near its shore.

Scientists who endured the dangers of diving at such a high altitude found it was well worth the risks. They were rewarded by being able to swim among the ruins of ancient buildings, wharves, and well-defined pathways. The quake must have hit with extreme suddenness;

evidence indicates that a number of small boats were still tied to the docks when they went down. Their owners never really tried to salvage them. Perhaps they felt the gods wanted the boats more than they did.

No timepieces marked this moment of passing, but the remains are from a civilization well over 2,000 years old.

Wise and Not-So-Wise Men of the East

The people of India have always been a source of wonder to the western world, especially in earlier times when primitive communication and travel kept the country mysteriously remote. Possibly because, within much of the religious thought of India, people can do very little to improve their expectations for this life, many aspire to transcend their physical existence however they can. They search for a mystical experience and use the power of inward concentration to arrive at supreme understanding and supernatural physical power. They may try, for example, to defy gravity, to feel no pain, or to walk on water.

The Mystic

In Bombay, India, in the year 1966 one such mystic had so convinced faithful followers of his prodigious skill that he himself began to believe that he was all-powerful. This Hindu yogi, Rao by name, calmly announced his intention to walk on water. Although none of his followers had ever witnessed a demonstration of his skill, they had no doubt that he was capable.

Arrangements for a public display of this talent that had been bestowed on him by the Most Holy One were made quickly and on a grand scale. More than 600 prominent members of Bombay society were invited to witness the spectacle, for a fee of course. Tickets, in great demand, sold for as high as $100 each.

On the day of the holy event, the assembled crowd was hushed and tense. The yogi, bedecked with a snow-white beard and garbed in flowing robes that fluttered gently in the breeze, stood on a rock overlooking the side of a pond five feet in depth. He posed so piously and majestically that all awaited the miracle in eloquent silence: blinking eyes or dropping pins would have interrupted the tranquillity. Praying inaudibly, Rao the yogi stepped onto the water—and promptly sank to the bottom of the pond.

Following Rao's dismal performance, mystics of remarkable self-proclaimed powers were greeted with some skepticism for a number of years in Bombay. But eventually people forgot Rao's folly in their desire to believe the unbelievable, and he has been replaced many times since.

The Debtor

As anyone with a charge account can attest, a person who owes a specific amount of money is given a certain amount of time to pay off this debt. In 19th-century India, as now, occasionally debtors down on their luck were not in a financial position to fulfill their obligation. Whereas today's deadbeat receives an emphatic overdue statement, the creditor in India went a different route. He would have the debtor seated in a public square, inside a circle drawn on the ground. Guards were posted around the debtor with strict orders to keep him confined within the circle. If he made the slightest attempt to cross the circle line, he was clubbed back into his public prison. If he dared to attempt an escape, he was immediately put to death.

Relatives were permitted to bring the debtor food and drink, but all necessities of private life were carried on in the circle before anyone who wished to stand by and watch. Passersby were usually unsympathetic and would jeer and scoff at the debtor's misery. No record exists of anyone's having perished in this open-air prison, because either he or some relative would come forth with the money owed from some "forgotten" fund. Needless to say, freeloaders were almost unknown in India; anyone unable to pay off a debt would head for parts unknown long before the due date.

The Snake Charmer

Another Indian attraction with mystical implication is the ability of some Hindu musicians to exercise control over cobras. As the Hindu fakir plays on his flute, the snake will sway side to side in whatever direction the musician indicates, as if the snake is being charmed by the music. Actually the cobra's hearing is extremely limited, so it is unlikely to respond to the music coming from the charmer's pipe. The cobras used by Hindu snake charmers actually respond to vibrations and movement: the tap of the charmer's foot, the beat of his stick on their basket, mainly the swaying of the musician's body and pipe. The rhythmic movements appear to fascinate the snake, so it will sway in a similar way, quite possibly looking for a place to strike.

Some of the more cautious snake charmers will remove the poison-inflicting teeth from the cobra before a series of performances. For a week or so the snake will scarcely attempt to chew on anything, including the charmer, because its jaws are too sore. Eventually it will be able to strike, but with its injectors missing, its bite is harmless. Snake charmers know that missing fangs regenerate and the process must be repeated. These repeated trips to the dentist have a negative effect on the cobra's health, shortening its life span considerably.

Surely there must be a better way for both snake and charmer to make a living.

A Faithless Wife

Much of the raven's social behavior appears to have been learned from humans. At age two, when two ravens enter into a committed relationship, they are strictly monogamous. Once paired off they practice lifetime fidelity. Since they live almost as long as humans, ravens celebrating their golden anniversary may look forward to another dozen years or more of togetherness.

A raven, just as its human counterpart, may occasionally stray. A scientist engaged in studying the home life of a pair of ravens observed this rare occurrence from start to finish. A newly mated female was being caressed by a strange raven while the husband was away from home. He arrived unexpectedly back at the nest and witnessed his faithless wife at play. He flew furiously at the intruder and drove him off. Again like a human husband, the offended male raven flew off and didn't return until late that evening. When he rejoined his repentant mate, all was forgiven.

The Most Dangerous Animals in Africa

Among all wild herbivores (plant eaters), the African elephant is the animal most likely to regard humans only with curiosity unless it perceives them as a threat. So there is little doubt now that fear and hatred of the hunter can be passed on from generation to generation.

For centuries elephants were hunted with poisoned spears or were driven into traps where they were impaled on poisoned stakes. By the end of the 19th century, hunted seriously for their ivory, an estimated 100,000 elephants were killed annually. The noblest of land creatures was being massacred for cameos and figurines, billiard balls and piano keys.

During the first 10 years of their lives young elephants are relatively immature and continue to learn about the world and how to survive in it from their mothers. Normally the African elephant entertains an extraordinary horror of the human hunter. Even the awareness of a human child windward of them can put a herd of elephants to flight. That an elephant can trample a hunter has no value against human weaponry.

How rapidly these perceptive animals become aware of a hunter's presence also is not surprising. All the elephants in a district are alerted to a hunter within two or three days. The elephants' method of communication is not entirely understood, but animal behaviorists have some ideas. They have observed, for instance, that hunters approaching an elephant herd often see one of the elephants raise its trunk, emitting no audible sound. Yet the herd is instantly alerted as far as a mile away. Actually the elephant did emit a sound, one of such low frequency that humans cannot detect it. Elephants may use this undetectable sound to communicate a warning to each other that human hunters are approaching.

When an elephant has nowhere to flee, it may react quite differently from the cautious, furtive beast that avoids confrontation. A cornered animal may become a most dangerous beast of prey. In the region now preserved as the Addo Elephant National Park in South Africa is a large, dense forest. Years ago the elephant herd that lived on the outskirts of the forest took to foraging in adjacent farms and orchards. Desperate over the devastation of their crops, the farmers hired a professional hunter to eradicate the destructive creatures.

Beginning in 1919 and over a period of months the hunter picked off elephants whenever and wherever he could get a shot at them.

Within a year the original population of 140 animals had been reduced to fewer than three dozen. The survivors were becoming increasingly difficult to find because they had retreated into the deep recesses of the forest. They were so wary of the hunter that they would emerge from the depths of the forest only at night. Then, every time the hunter could get close to them, one of the elephants would charge at him from dense cover. They appeared to wait in ambush for their avowed enemy. After several very narrow escapes the hunter gave up and resigned from his position as elephant eradicator.

Although the descendants of these elephants are now protected by strict game laws, the herds have not forgotten their former persecution. Unlike most African elephants they still come out from the protective cover of the forest to feed only after dark and will charge any human on sight. This small herd of elephants is currently considered, by authorities and poachers alike, to be the most dangerous group of wild animals in Africa.

The facts of life that the elephant child learns at its mother's knee are not forgotten.

Ravishing Raisin

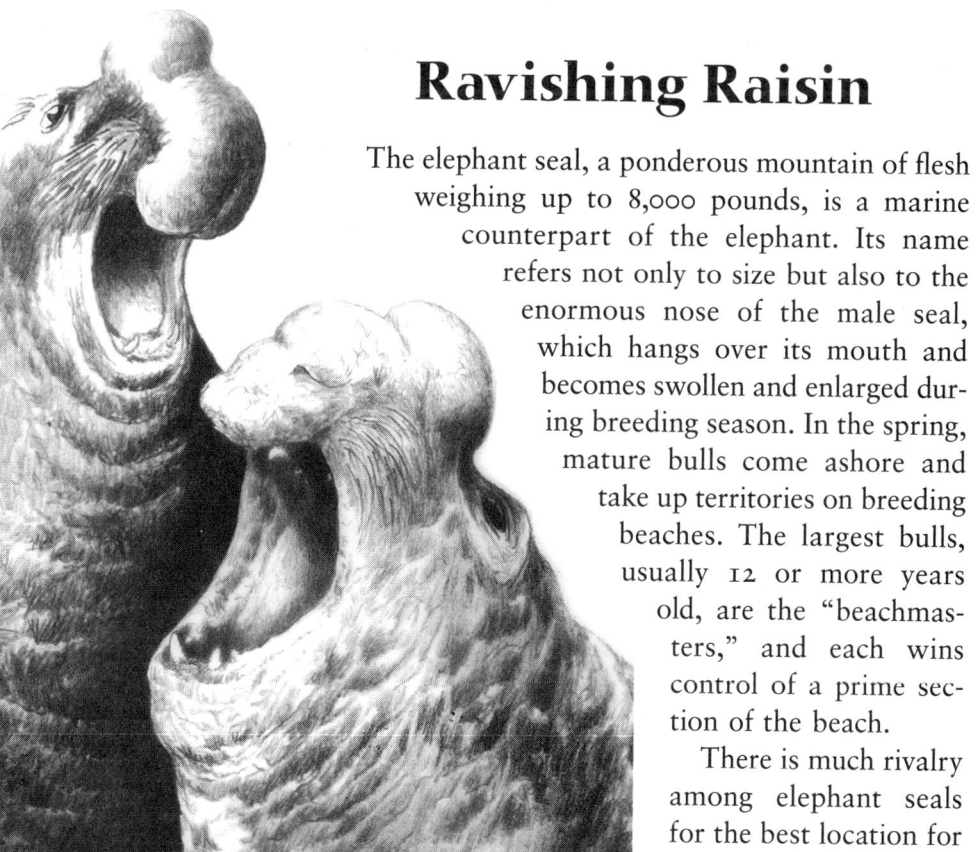

The elephant seal, a ponderous mountain of flesh weighing up to 8,000 pounds, is a marine counterpart of the elephant. Its name refers not only to size but also to the enormous nose of the male seal, which hangs over its mouth and becomes swollen and enlarged during breeding season. In the spring, mature bulls come ashore and take up territories on breeding beaches. The largest bulls, usually 12 or more years old, are the "beachmasters," and each wins control of a prime section of the beach.

There is much rivalry among elephant seals for the best location for breeding. The huge males constantly challenge each other with loud roars that can be heard for several miles. As a bull snorts, its oversize nose directs the sound waves into its mouth, which acts as a resonance chamber (the proverbial bullhorn?). Many vicious but rarely fatal fights break out. Rarely does an elephant seal not bear the scars of previous encounters.

When the females finally arrive on the beach, they all tend to ignore the males. They are preoccupied with giving birth to pups, after a gestation period of 350 days. A newborn seal can weigh 100 pounds, and its rate of growth is remarkable but understandable. Feeding from its mother's milk, which is 5.5 percent fat, it grows an average of 10 pounds a day. The pups are cared for until they have quadrupled their weight—a mere four weeks. They will then be weaned from mother and join other weaned young seals in groups called weaner pods. The nursing mother never leaves to feed, and her fast can cost her as much as 500 pounds in body weight, trimming her quite a bit from her original 1,600 pounds. When the pups are on their own and seaworthy, the female is ready to breed again.

Because no male seal will leave his breeding place, he cannot feed during the three-month mating period. Meals of his favorite fish and squid, found at a depth of 300 feet, are not available to a seal protecting his section of beach and a harem. He spends his hours and energy running off rival males and impregnating dozens to hundreds of females he has rounded up for just that purpose.

Scientists were curious about just how much energy the male seal invests in this three-month period of fasting while engaged in very energy-consuming activities. One biologist felt that calculating weight loss during the breeding season would be a good index of the reproductive investment the male seal puts into this exhausting activity. The problem was to find a way to get a hefty male elephant seal onto a scale during his busily obsessed season. The answer became clear: take advantage of his well-known libido. Drugging the male during a time of such high-energy preoccupation could kill him, so the scientist planned to lure the love-intoxicated bull with a female facsimile. She was a model of a seal mounted on a surf-board base and affectionately named Raisin.

The imitation seal was extremely well constructed, with appropriate appearance, sounds, smells, and movements. Raisin's tail was automated, and her wagging was synchronized with recorded sounds of an actual female being mounted by a male. Raisin was placed just beyond a sand-covered sensitive scale.

A young male named Cod 4 was the first to fall for the ruse, and he eagerly rushed onto the scale to get to Raisin. A scientist concealed in shrubbery read off "1,919 pounds." By the end of the breeding season

Cod 4 had lost more than one-third of his body weight, and, counter-productively, had been unable to copulate with any of the 350 females in his harem. His weight was lost by the combination of fasting and chasing amorous males away from his breeding area as they attempted to steal some of his females. The harem-nappers were often successful because while Cod 4 was busily chasing one male another would rush in and drive off one or more females. So preoccupied was Cod 4 with guarding his wives that he could find no opportunity to relax and propagate his race.

Not all the males on the beach fell into so unrewarding a situation. The most dominant male on the beach, a monstrous seal named Outer by the scientists, managed to inseminate 87 females during the mating season. He too paid a heavy price. Early in the season Outer had also fallen for the charms of Raisin, so the scientists had an opportunity to record his original weight: he tipped the scale at over 5,000 pounds. By the end of the season Outer had lost nearly half of his original weight by fasting and protecting his harem from over 50 marauding bulls.

Many of the males looked with favor upon Raisin, and quite a few tried to mount her. All experienced extreme frustration, and none made any further effort to add her to their harem. Although they may not have noticed, Raisin didn't seem to care.

When the mating season is over both male and female seals will return to the sea to feed. During this period each new embryonic seal carried by its mother is in an arrested state of development. Actually it's nothing more than a hollow ball of cells floating free in the womb. In a process called delayed implantation, the embryo's development is stopped for several months while the mother regains the weight she lost. Once the female recovers, the ball of cells implants and growth continues.

About 11 months after pregnancy began, the mother seal will return to the beach where she mated almost a year ago and give birth to the new offspring. She will then mate again, perhaps this time without competition from Raisin.

Symbolic Barber Pole

An elderly acquaintance recalls that, during his youth, whenever he complained of a toothache his father would send him to the corner barbershop. For 50 cents the barber would (without anesthesia) extract the defective tooth. Another senior citizen will never forget a most embarrassing moment during his early teens, when his mother pointed out to the barber the all-too-obvious acne on his face. Knowing that barbers are enlightened in such medical matters, she asked him what he would recommend for her son's pimples. The barber, eager to justify Mom's faith in his expertise, tilted the young man's face to the light, scrutinized it thoughtfully, and prescribed in a fine, authoritative accent, "Use-a lemon juice."

Evidently barbers, whose craft had been intermingled with that of surgeons since about A.D. 500, were still practicing on-the-spot medicine in the 20th century. The partnership was not always based on mutual respect, and much of the time few regulations governed who was responsible for what particular treatments. As early as the 10th century, Albucasis, the great Arab surgeon and author of the first illustrated book of surgery, cautioned dental surgeons to avoid acting like ignorant and foolish barbers. He complained that they caused patients great injuries and were often guilty of removing the wrong tooth, breaking a tooth, leaving the root in the socket, and other inexcusable oversights.

Barbers became important figures in monasteries especially after 1092, when beards were banned and each monastic order had its special hairstyle. With the Edict of Tours in 1163 that held it to be sacrilegious for clergy to draw blood, monks were forbidden to perform operations, and surgical procedures fell to the barbers who had been their assistants. Separated from the clergy, barber-surgeons continued to practice for several centuries, expanding the scope of their practice somewhat. In the 13th century, along with bloodletting, tooth pulling, and cauterization, the craft-oriented hierarchy of barber-surgeons also became the suppliers of heat treatments and physical manipulation to set fractures. Surgeons were identified, not from university training as were the physicians, but from apprenticeship within the guilds.

In the 14th century some restrictions were put on those surgeons whose skill was an outgrowth of shaves and haircuts. But by then bloodletting, cupping, leeching, giving enemas, and extracting teeth had become the almost exclusive province of the barber-surgeons. Because bloodletting was an alleged panacea for so many ills, it was the treatment to which many ailing people would submit. Bloodletting equipment included a basin to collect the blood and a pole for the patient to squeeze so that veins would swell and blood would gush freely. The pole was painted red to conceal stains of blood overgushing and later was wrapped with white gauze to suggest the bandage applied to the blood-let arm. Hung outside the shop to advertise the service, the pole became a trademark of the barber-surgeon guilds. The red and white pole was certainly more tactful than the buckets of blood and bloodstained rags previously displayed to advertise the medical services of barbers.

One of the most notable surgeons who arose from the ranks of barber's apprentice was Ambroise Paré (1510–1590). He became a military surgeon to extend his operating skills and made several momentous discoveries. The established practice for the treatment of gunshot wounds was to cauterize them with boiling oil. When, after an extremely bloody battle, the supply of oil ran out, Paré was obliged to fabricate his own wound dressing. He covered the wounds with an unguent of egg whites, oil of roses, and turpentine. The next morning he found the soldiers thus treated were comfortable, while those who had been treated with boiling oil were feverish and in great pain. He resolved never to "cruelly burn poor, wounded men." Paré later became famous as a surgeon to several French kings. This barber-surgeon is known as the father of surgery.

Over several centuries most barber-surgeons continued with lucrative medical procedures to supplement the tonsorial business, which rose and fell as fashion dictated. Eventually surgeons and barbers parted ways, and the barbers inherited the pole of the barber-surgeon guild. The rotating red and white striped pole has continued to designate barbershops up to the present, although a pole of patriotic red, white, and blue is now more common. In earlier days the absence of a barber pole seemed to suggest that a fly-by-night nonprofessional was wielding the clippers, scissors, and razor.

Today, barbershops have again extended their services far beyond shaves and haircuts, but in the area of grooming rather than surgery. Most provide services for both men and women, and hairstyling has replaced the archaic "one-style-fits-all" haircut. As an easily recognizable piece of nostalgia, the elegant, unpretentious barber pole should remain a distinctive symbol in our cities. Like the three gold balls over the pawnshop, the barber pole refuses to become homogenized.

Rise and Fall of a Hurricane

*People caught in a hurricane's fury often felt
the earth was in its death throes.*

A hurricane is the offspring of the tropical seas, a whirling mass of energy formed when certain conditions of heat and pressure nourish and nudge the winds over a large area of ocean. The winds wrap themselves around an atmospheric low (the eye) with a significant degree of high energy. The hurricane is a heat engine generated and maintained by the energy of condensation of rain within its saturated spiral clouds. It is nature's way of releasing heat that builds up in the tropics.

The word *hurricane* is derived from the name *Hunraken*, an ancient Mayan storm god; among aboriginal Caribbean people a similar word, *Hurakán*, was the name of an evil spirit. In the North Pacific this same storm is called a typhoon (from the Cantonese *t'ai fung*, meaning "great wind"), and in the South Pacific and Indian Oceans it is a "tropical cyclone."

Witnesses who survive an encounter with a hurricane describe a most unforgettable experience. With the hurricane blowing in your face you can't breathe, nor can you see with rain and spray hitting your eyeballs at 100 miles per hour. You hear only the screaming of the wind that drowns out even thunder and roaring breakers. Speaking is impossible with your mouth blown out of shape, and if you try to walk or crawl you will be blown away. You can only hope to happen on to something solid around which to twist arms and legs.

Hurricanes form over warm ocean waters, 80 degrees Fahrenheit or higher, usually 5 degrees to 30 degrees north or south of the equator. The first and most important condition necessary for the formation of a hurricane is a vast stretch of open sea that has been heated decisively by the sun beaming down day after day from a cloudless sky. The warm sea creates an air funnel in which hot, moist air rises and condenses; heat is released, and the action increases. This funnel of warm air rises to perhaps 40,000 feet, producing vast cumulus clouds and violent thunderstorms. While more and more moisture is drawn from below into the funnel, these high air currents are distributed to the surrounding areas. The earth's rotation gives the system a twisting motion that gathers momentum, and a hurricane is born. It whirls counterclockwise in the Northern hemisphere, and clockwise in the Southern.

The hurricane is like a swirling doughnut with a hole in the middle, with winds over 75 miles per hour, averaging 130 mph, and reaching 200 mph or higher. As the storm system grows it may reach 400 to 500 miles in diameter. It is nature's most destructive storm because it covers such a wide area, traveling thousands of miles and raging for a relatively long time, usually about nine days. The San Ciriaco storm of 1899 churned the Atlantic waters for almost five weeks. The average traveling speed of an Atlantic hurricane is 12 mph, although it varies widely; a hurricane in September 1938 traveled through New England at about a mile per minute.

The energy involved in a hurricane is mind-boggling. The typical hurricane liberates heat from the condensation of moisture at a rate of 100 billion kilowatts per hour. This means that in a single day the average hurricane releases more energy than the electricity used by the entire United States in an entire year. Or, compared with nuclear power, the energy released by a hurricane in one day is equivalent to the explosion of 2,000 one-megaton hydrogen bombs. In addition to energy, the average hurricane squeezes out an estimated 200 billion tons of water a day as rain!

The direct action of the wind is not the most colossal wreaker of havoc in a hurricane. Most victims die by drowning, and most of the flooding comes not from the intense rains but from the storm surge. Winds and low pressure around the hurricane eye raise the level of the sea several feet. This produces a dome of water about 50 miles across that moves onto the coast, bringing the seas far inland. The National Weather Service estimates that storm surges cause 9 of every 10 hurricane fatalities.

The maelstrom of winds moves in a vast spiral around the storm center—the eye, the hole in the whirling doughnut. The eye is an oasis of comparative calm and, occasionally, sunshine. The temperature is appreciably higher than in the surrounding vortex. This is the low-pressure area around which the furious winds revolve. These winds become more violent—100 to 200 mph—at the approach to the calm center. Suddenly the eye arrives, and there is an almost instantaneous lull. Winds may drop to as low as 10 mph, rain stops abruptly, and patches of open sky are visible above. The calm continues while the eye passes; its average diameter is 14 miles, but it can range from 4 to 25 miles. This is only a brief time-out for battered victims of a hurricane; as soon as the eye passes, the destructive winds resume but from the opposite direction.

When the eye passes over a ship, flocks of birds land on the deck and rigging, sometimes completely covering all surfaces. The pilots of a weather plane observing Hurricane Carla in September 1961 reported the air so full of birds that they dared not fly through it. Any bird caught up in the violent winds is powerless and is carried willy-nilly wherever the winds flow. As the winds spiral toward the center, the birds eventually reach the calm eye, and there they must remain until the hurricane subsides. When the eye passes over a ship, the birds will land on it to rest from their life-and-death struggle with the elements. These birds sometimes travel great distances in the eye. Tropical birds have been found in New England, 2,000 miles from where they "boarded" the hurricane.

At sea the storm's energy is replenished by the constant drawing of warm, evaporated water into the funnel. But once the storm moves over the mainland the supply of evaporated water from the ocean is cut off, and its power eventually diminishes. Then the hurricane slowly dies, but not before it dumps enormous quantities of rain on the land below. Hurricane Diana (1955) caused little damage along the coast, but the rains brought floods to New England and the Mid-Atlantic, killing over 200 people and causing about $300 million additional damage.

The relentless violence a hurricane can inflict may inspire fear and dread, but its silver lining is the accompanying rainfall. Although the deluge may be too much, too fast, the average hurricane drops six inches of water over a given area. As parched land is revived, rivers resume their normal flow, and reservoirs are filled. Drought-stricken plants and animals are refreshed. Hurricanes, typhoons, or tropical cyclones are essential to successful agriculture in many countries. Toilers of the land depend on these storms for one-fourth of needed rainfall, and nature is more than willing to oblige.

More than a Mouthful

Of the many ways that animal species protect their young, the mouthbreeder has the most foolproof system. As its common name suggests, the female fish of the genus *Haplochromis* incubates the eggs in her mouth. Even more unusual, she takes the eggs in her mouth the moment they are laid, before they have been fertilized by the male.

This would be a monumental oversight except that the male is equipped to deal with it. The anal fin of the male mouthbreeder is sprinkled with bright orange spots that look very much like the mouthful of eggs the female is carrying. The sight of these spots inspires her to try to add them to her brood. As she opens her mouth she takes in milt (sperm-laden fluid) that the male releases into the water, and fertilization takes place.

The mouthbreeder refuses food during the two to four weeks that the brood is being incubated in her mouth. As the eggs grow, her lower jaw stretches and becomes deeper and her belly shrinks, so she appears to be all head. When tiny eyes can be seen on the eggs in her mouth, she is about to give birth. About 30 to 80 are expelled in a stream, and the mother swims near them for a week or more. Whenever danger threatens they jostle each other in a frantic attempt to swim back to the safety of their mother's mouth. If they should be snapped up by any other adult mouthbreeder they will be eaten.

The mouthbreeder is one of the family of cichlids, many of which are popular tropical fish for the aquarium. Fanciers who wish to witness the incubation of eggs in a fish's mouth and watch the fry seek refuge in their mother's jaws will enjoy the mouthbreeder. The experience will be worth the trouble of placing aggressive fish in a separate tank or removing a fish with a brood from the threat of hungrier fish.

CHAPTER EIGHT

A Fraud Called *Missourium*

In 1841 a curious individual by the name of Albert Koch placed on exhibition a rather strange fossil skeleton of a creature that superficially resembled an elephant but made ordinary mammoths and mastodons look like pygmies. He called the creature *Missourium*.

Who was Albert Koch? We know little about him and his early life. Koch was born in Germany and came to the United States about 1835. What scientific training, if any, he had was completely unknown. He always referred to himself as "Dr." Albert Koch but never explained where or how he had acquired his degree.

Koch settled in St. Louis, where he opened a business as a dealer in fossil specimens. He went on frequent fossil collecting trips through many parts of the southern United States, tracking down reports of fossil finds and persuading farmers to let him dig on their land. His most spectacular fossil finds he placed on exhibit in a sort of traveling circus; minor pieces and scattered bones he sold to museums or universities. Any exhibited, mounted fossil that lost public interest was quickly offered for sale.

The man did not behave like a scientist, and for this he was deeply despised by real scientists. They looked on him, quite accurately, as a peddler of old bones and a charlatan showman making a profit out of relics of the past. But despite the fact that his approach to scientific specimens was strongly commercial, he did dig up a great many specimens that could have been of value.

Early in 1840 Dr. Koch unearthed a nearly complete skeleton of a mastodon in Benton County, Missouri. It lay on the shore of the small Pomme de Terre River, covered over with clay, gravel, and quicksand. This was the second time such a nearly complete mastodon skeleton had been discovered, and thus it was an important scientific event. But Koch handled it in his own special way to make it seem even more important than it really was.

Later in 1840 Koch began to mount the Benton County skeleton for exhibition. It was a fine specimen of *Mastodon americanus*, but Koch improved on it quite a bit by weaving the bones of other mastodons into his exhibit. He added ribs and vertebrae until he had created a true monster. Out of three mastodons, Koch pieced together the *Missourium*, 32 feet long and 15 feet high. He completed his masterpiece by attaching the tusks to the skull in such a manner that they curved up and back over the animal's head like a pair of gigantic horns, instead of jutting downward and outward in the proper fashion.

The finished product was awe inspiring, and Koch hauled it all over the country on a highly profitable exhibition tour. On October 15, 1841, in Philadelphia, the American Philosophical Society heard a speech by Dr. Richard Harlan, one of the United States' leading authorities on fossils. "There is now exhibiting at the Masonic Hall in Philadelphia, one of the most extensive and remarkable collections of fossil bones of extinct mammals which have hitherto been brought to light in this country." Dr. Harlan was great in his praise of Dr. Koch and was surprisingly gentle in dealing with the grotesque distortions and errors Koch had made in assembling the skeleton of *Missourium*. Of such things as attaching the tusks upside down to form horns, Harlan said merely that "no doubt" Koch's further researches "would enable him to rectify these."

By the end of 1841 the majestic *Missourium* went on display at the Egyptian Hall in Piccadilly, London. Such an exhibition had never been seen before in England. It created quite a sensation; immense crowds came to view it, and the money rolled in.

Sir Richard Owen, England's foremost anatomist and paleontologist, came to view the display and eyed the monster with suspicion. It seemed to have far too many ribs, and the horns looked like upside-down elephant tusks (which they were). Owen, on February 23, 1842, read a paper on Koch's *Missourium* to a meeting of the Geological Society of London. He disputed the exhibit as a mastodon that had been mounted incorrectly. Koch had the audacity to challenge the great Englishman on his home grounds; on April 6, he addressed the same Geological Society and insisted that he had indeed unearthed a previously unknown genus.

The debate did not seem to affect business. Month after month the customers thronged into Egyptian Hall, and the money kept coming.

In summer 1843 Koch decided to close his London show and went on tour with his monstrosity. He first went to Ireland and then to Germany. In the latter country his fossil aroused even greater interest than it did in England. Customers were never lacking, and the money continued to roll in.

Koch returned to the United States in May 1844, but with a brief stopover in England, where he sold *Missourium* to the British Museum. He received an extraordinary price for his mastodon: $2,000 down and $1,000 a year for the remainder of his life. Evidently the museum officials didn't expect him to live long, but he hung on long enough to receive $23,000 before his death in 1866. Ironically the scientists at the museum knew the fossil was a fraud. As soon at the specimen arrived, the paleontologists took the monster apart, stripping away the extra bones and putting the tusks where they belonged. The dehorned skeleton, correctly labeled *Mastodon americanus*, is still to be seen at the gallery of fossil mammals of the British Museum's natural history collection at South Kensington. It is still one of the finest mastodon fossils in existence.

Almost immediately after he sold his monstrosity in 1844, Koch showed up with a new wonder: a gigantic 114-foot-long skeleton of a "sea serpent" he called *Hydrarchus*. It also drew tremendous crowds in American exhibition halls, and again the money rolled in. But a famed zoologist named Jeffries Wyman soon pointed out that *Hydrarchus* was actually the fossil of an extinct whale that Koch had decorated with bones from at least four other animals. Wyman was right on!

Koch immediately closed up his U.S. tour and left for London, where he received a rather chilly reception. So he and *Hydrarchus* were off to Germany, where he was well received. He rewarded the Germans by selling them *Hydrarchus* for a fine price. Once again the money rolled in.

In 1838, two years before the good doctor presented his *Missourium*, Koch appears to have been more scientifically inclined. He actually discovered the remains of a mastodon that was clearly associated with artifacts of early man. Plenty of evidence supported his find, and Koch wrote a report on it, which he published in a Philadelphia magazine called *The Presbyterian*. He did an excellent job describing his find, but since nothing like this had been found in the United States, belief was slow in coming. Too many people doubted that human beings had been around to hunt "antediluvian" creatures. Other than that one article, Koch kept his find quiet. But in 1841, when he was in Kentucky exhibiting his *Missourium* concoction, he issued a booklet describing his various fossil finds. The book included a section on early human hunts for mastodons and gave an account of what must have happened. The hunters came upon a disabled mastodon and killed and roasted it on the spot. Koch went on to describe in great detail the numerous artifacts found with the mastodon.

These really were important discoveries. They showed that humans had lived in the New World for many centuries. But few scientists took Koch seriously; because he proved to be an archeological profiteer his findings were wholly discredited. He finally settled in St. Louis and eagerly sought to win scientific respectability.

That he was regarded as a fraud by the men whose esteem he most dearly craved really seemed to bother Koch. He lived quietly in the last years of his life, publishing no more and harboring deep bitterness toward those who refused to accept his one genuine contribution to science. True, he had put horns on a mastodon, then doubled the length of a whale and called it a fossil sea serpent. He had committed many a crime against truth. But he had also shown that humans in North America had hunted prehistoric pachyderms, and it pained him that his fondness for fame as a showman had robbed him of credit for that scientific achievement.

The Hoax That Was Real

In 1880 a stuffed specimen of an animal arrived in England from Australia. Australia had a reputation (well deserved) for being the source of the most unusual animals and plants in the world. But this new arrival baffled everyone and raised quite a few eyebrows.

The specimen had a thick coat of hair, but it also had a bill and webbed feet one would expect to find only on a duck or some related species. The animal also had a spur on each hind ankle for secreting poison, as one would expect from a snake or related reptile. Oh, and later, when a female was examined, no poison spur was found, but under the tail was a single opening that was definitely used for laying eggs.

As one might expect, zoologists of the day immediately suspected a hoax, that this specimen had been artificially put together with parts from several unrelated animals. However, a close examination by scientists could find no evidence whatsoever of any artificial joining of any body parts. Experts were baffled by this new arrival from the land down under. What could this combination mammal-bird-reptile possibly be? Well, this animal, which in life lives both on land and in water, has since been intensely studied. Of course it was no hoax. Europe was just not ready to accept the duck-billed platypus, a true mammal that lays eggs and nurtures its offspring with milk as any other mammal does.

Mesozoic Lifestyles

Prehistoric remains often tell amazing stories of how an animal lived and died. Among the more interesting are those that help to interpret the lifestyles of the many dinosaur species and their associates that lived from 250 million to 65 million years ago during the Mesozoic Era.

Crocodiles lived side by side with dinosaurs that played a large part in the daily crocodilian menu, as shown by the following tale recreated from their remains. During the early 1960s two 30-foot crocodile skeletons of the genus *Parasuchus* were found in an ancient floodplain deposit in central India. The remains were less than a meter apart and lying at such an angle that the crocodiles appeared to be in communication when they died. One may have been complaining to the other of the excruciating stomachache it was having, and the other, experiencing the same agony, understood completely. Their discomfort could have been related to the fact that within the rib cages of both skeletons of *Parasuchus* were the skeletons of the four-foot dinosaur Malerisaurus. Both of the Malerisaurus appeared to have been gobbled up at the same time.

From the evidence of the skeletons and the surrounding sediments, the scientists were able to reconstruct what had happened. The two

dinosaurs were drinking side by side at the edge of a stream, unaware of the crocodiles lying submerged and preparing to strike. The predators grabbed the victims simultaneously, pulling them into the water, where they drowned. Four-foot dinosaurs were tidbits to the crocodiles, and they were swallowed whole almost immediately.

The most probable explanation for the simultaneous deaths of the crocodiles is that the skin of their victims was poisonous. Their Malerisaurus meals were followed by some agonizing moments for the *Parasuchus*. They writhed in torment briefly, twisting and turning, possibly even snapping at each other until the relentless pain was quieted by their death. They gradually sank into the soft bottom muds. Here the victims lay in close proximity, as if each were describing and comparing its demise the way it happened nearly 180 million years ago.

The most popular members of the dinosaur clan have always been the sauropods. No child's treatment of these ancient creatures would be complete without the lovable *Brontosaurus* (*Apatosaurus*), an almost generic dinosaur. The sauropods were by far the largest animals ever to walk the earth. As one paleontologist observed, "Sauropods were beasts of the size of whales, on legs of elephants, with tails of lizards, necks of giraffes, heads of horses and nostrils of tapirs." They were among the most successful of any dinosaur group.

The dentition of the sauropod dinosaurs indicates that their diet consisted largely of marsh plants, although the largest ones undoubtedly browsed on leaves they nipped off of large trees with their blunt teeth. To swallow they probably had to raise their heads and let gravity help the neck muscles. How strange it would have been to watch a herd of feeding sauropods! Rhythmically raising and lowering their heads, they must have resembled a battery of dredges.

The most remarkable characteristic of the sauropod dinosaurs is their size. A recent contender for the largest known sauropod species was discovered in New Mexico in 1985 and is currently being excavated. It appears to have been, from the tip of its tail to the nose, about 140 feet long and weighed anywhere from 60 to 90 tons. Scientists named it *Seismosaurus* because the ground must have shook when this great beast walked. A footprint of a *Seismosaurus*, or a near relative, was found in the area of the Paluxy River in Texas during excavations over 50 years ago. A famous photo of the find shows a boy about six years old taking a bath in the fossil footprint. It was almost bathtub size because when completely filled it held over 18 gallons of water.

Scientists have contemplated, with difficulty, just how much vegetation a full-grown adult *Seismosaurus* would eat. The dinosaur must have fed almost constantly, its long neck serving as a funnel that led to the huge body reservoir. The undersized head was like a rake, ceaselessly scooping and swallowing whatever vegetation was within reach. It did not chew its food; instead the *Seismosaurus* swallowed enormous numbers of plum-sized stomach stones (gastroliths) that served as teeth and ground up the food as it passed into the primary stomach. In the rib cage of the skeleton being excavated in New Mexico, over 230 of these stomach stones were found, all quite similar in size. Near the rib cage, however, was a stone the size of a large grapefruit. Scientists studying the specimen believe this could have been a death rock. *Seismosaurus* may have deliberately swallowed an oversized rock to add some vigor to its collection of internal "teeth."

This rock seems to have been too large and lodged in the throat of the beast, causing it to choke to death. A perfect case for the Heimlich maneuver, which unfortunately for this dinosaur would not be in use for another 154 million years!

Tyrannosaurus rex, king of the tyrant lizards, is among the best known of the carnivorous dinosaurs. It stood 16 to 20 feet tall, was nearly 50 feet long, and weighed well over six tons. Its jaws, at least three feet in length, were armed with numerous serrated teeth six to seven inches long. Although bipedal (two-footed), with its body held in a horizontal position it ran very fast. Toothy jaws agape, roaring in anticipation, it must have been a living nightmare for its prey. Encounters would have been, in most cases, short lived, and *T. rex* would feed.

Tyrannosaurus rex, although the largest carnivore to have walked on the globe, was not guaranteed an easy life. Far from it. In August 1990 the largest, most complete skeleton of *T. rex* was found in South Dakota. The bones indicate that it was a female, so she became Sue.

During her lifetime Sue suffered a number of major injuries. She must have engaged in a deadly fight with another tyrannosaur because one rib harbors the point of a tooth within an abnormal growth of bone, showing that the bone had healed after the bite wound. A hole in the skull and a lump on the lower jaw also represent healed wounds, possibly from the same fight.

Both legs were broken during her violence-prone life. The two fibulae must have been fractured on separate occasions, because she couldn't have survived two broken legs at once. One can easily theorize how the breaks occurred. Scientists know that a charging *Tyrannosaurus rex* could reach speeds rivaling those of a modern racehorse. As Sue bore down on her prey she unwittingly stepped into a depression or a hidden hole in the ground and was pitched headlong on her face, with a leg dangling from the sudden break. She managed to survive this type of mishap twice! It was noted that several of Sue's tail vertebrae were fused together. Because this was a common injury among large female dinosaurs, it most likely occurred during the mating process.

Just before Sue died she appears to have dined on the plant-eating dinosaur *Edmontosaurus* as several bones of this dinosaur were identified in her rib cage and were etched as if by stomach acid.

Sue met her death in a fight with another tyrannosaur. The left side of her skull shows definite injury from a heavy bite, and there is no sign of recovery. Sue's antagonist doubtless succumbed along with her, for her skeleton shows no major signs of any chewing on the bones. Her opponent probably staggered off some distance before it too lay down for the last time, about 75 million years ago.

Although incomplete, the sad tale of *Therizinosaurus* (scythe reptile) is worthy of note here. A few years younger than Sue, *Therizinosaurus* passed away about 70 million years ago when it went to a lake to drink. This dinosaur may have resembled a present-day anteater, for it had enormous claws that would have been useful for ripping apart insect nests.

Because it had no way of knowing that a large carnivore lay in ambush near the watering spot, this specimen of *Therizinosaurus* never finished its drink of water. Instead it fell prey to the carnivore and was almost completely consumed. Its few remains include parts of the forelimbs and hind limbs, flattened ribs, a tooth, and several enormous claws. Scientists have been able to speculate very little about this dinosaur, but they agree that it must have been gigantic. One of its foreclaws, preserved in the Museum of Paleontology in Moscow, measures a staggering 28 inches along the outside curve.

The World's Worst Weather

The top of Mount Washington in New Hampshire has been designated as the place with the worst weather in the world. Situated at the convergence of three major storm pathways, the 6,288-foot summit of Mount Washington almost never has a nice day.

Combinations of bitter wind, intense cold, snow and ice, clouds, and fog are responsible for Mount Washington's unsavory reputation. The strongest wind ever measured on land, 231 miles per hour, was clocked on this mountain. About 250 inches of snow falls annually at its summit, and the average year-round temperature is below freezing. A layer of permanently frozen ground extends from a depth of about 20 feet to 100 feet below the surface. The mountain's official low temperature is 47 degrees below zero Fahrenheit, but cold combined with consistently high winds at the summit produces wind chill temperatures that often exceed 150 degrees below zero. On one out of three days between November and April, the winds exceed 100 miles per hour, and clockings of over 150 miles per hour are recorded almost every winter.

As inhospitable as Mount Washington is, the mountain has always held a fascination for adventurers. Called "Place of the Great Spirit" by native people of the region, the mountain had become a popular tourist attraction by the 1860s, when parties of mountain climbers flocked to the summit. Nearly a hundred people have died in their attempts to climb the mountain. Deaths resulted from falls, exposure, and simple inexperience in how to deal with the capricious mountain weather. No matter how bad the weather at the beginning of the climb, it can always take an impulsive turn for the worse.

One of the best known and earliest of all the mountain's recorded victims was Lizzie Bourne, a member of a climbing party that began an ascent to the peak on September 14, 1855. Arriving at a rest stop, Halfway House, the group was advised to stay overnight and complete their climb in the morning. Disregarding the advice, the climbers continued upward and quickly encountered strong winds, freezing cold, and dense fog. Finally the exhausted adventurers built a stone windbreak and huddled together to spend the night. Lizzie Bourne died of exposure in the late evening. The next morning, the survivors discovered that they had given up their struggle against the storm only a few hundred yards away from Tip Top House, an inn located at Mount Washington's forbidding summit.

Termite Soldiers

In the attempt to protect themselves, termites know that the best offense is a good defense. Their first, best defense is the safety of numbers, for a colony of termites can comprise as many as 10 million individuals. Another exceptional defense is their termitarium home, efficiently ventilated and air-conditioned to maintain a suitable temperature and humidity in all seasons. The nest is built of recycled materials such as pellets of soil, grains of sand, and particles of wood, all bound together by termite excreta and saliva. It has overhanging roofs to keep it rainproof and a formidable outer layer for protection. Only the claws of aardvarks and similar ant-eating specialists are able to break open a termite mound far enough to insert their long, slender, sticky tongues.

Another important termite defense is the soldier termites. Any colony has several castes of soldiers whose bodies are built specifically for defense. They are on duty at all times and defend to the death. The average soldier termite is easily identifiable by its distinctive body characteristics: oversized heads, powerful jaws, and usually black color.

Termite soldiers are specialists at fighting their archenemies, the ants. When an army of ants attacks the nest, they cut a hole in the wall of the termitarium. The termite soldiers immediately protect the breach and tear the invaders to pieces with their powerful jaws. They seal the hole temporarily by blocking it with the large head of a soldier. This often means the sacrificing soldier loses its head, but the invaders are denied access to the nest while the termites repair it permanently.

Several termite species have a unique, even more specialized, group of soldiers. They are called nasutes (which means having a large nose) because their foreheads extend into a nose-shaped point. When an invader breaks into the termitarium, the nasute soldiers gather at the threatened point. From the long beak a jet of fluid is expelled at the intruder. The secretion is not only toxic but also sticky. An insect enemy covered with it is rendered totally helpless.

Termite soldiers appear invincible with their built-in weapons, powerful jaws that can snap an ant in two, and effective armament. All of this is necessary because termites are no match for ants if the two face each other in open battle. Careful planning and extraordinary defenses of their nests are the keys to their survival.

The Slave and the Milkmaid

For many centuries a dreaded infectious disease called smallpox ravaged Europe and America. Each epidemic killed about 40 percent of those who contracted the disease, which permanently disfigured all its victims. Now, in the final decades of the 20th century, smallpox no longer presents a threat. For the discovery of that medical milestone, civilization is indebted to an African slave in Boston who belonged to Cotton Mather.

Mather (1663–1728) was a Puritan minister whose favorite slave was named Onesimus. When Mather observed that Onesimus seemed to be immune to smallpox, he wanted to know why. Onesimus explained his immunity to Mather but was not sure that he would believe the story.

Onesimus told of a procedure used commonly in his homeland in western Africa. A drop of liquid from a smallpox sore was put into a cut on a healthy person's arm. That person would often become slightly ill but would rarely, if ever, get smallpox. The small drop of liquid containing a minute amount of smallpox virus rendered the person who received it permanently immune to the dreaded disease.

Mather told his medical friends about the procedure, a forerunner to vaccination. Most were unwilling to accept the idea and considered it dangerous to purposely infect a person with smallpox. Mather persisted, amid much resistance, until several medical practitioners began to use the procedure described by Onesimus. The results spoke for themselves, as those who were vaccinated became immune to smallpox.

During the 18th century, immunization against smallpox scarcely moved from its Boston birthplace. Not until the 1770s did the English physician Edward Jenner note that milkmaids exposed to cowpox developed nodules on their hands but did not get smallpox. He successfully immunized a boy with pus from a nodule on a milkmaid's hand and then continued to experiment in the face of controversy. He was eventually able to provide vaccine and encourage vaccination throughout Britain.

In 1801 Jenner predicted that vaccination would eradicate smallpox worldwide. In 1980, just 179 years after Jenner's prediction and 274 years after Onesimus told his master how he happened to be immune to smallpox, the World Health Organization made a most dramatic announcement. Because of a successful worldwide vaccination campaign, smallpox was now extinct.

Creatures That Time Forgot: Living Fossils

Rumors of dinosaurs, sea serpents, and other fantasies appearing in remote corners of the globe crop up frequently in newspapers, magazines, and especially tabloids. No one can squelch the hope that dinosaurs still cling to life on some unexplored island; the idea that ancient forms still exist appeals to almost everyone. This hope explains the popularity of Arthur Conan Doyle's 1912 classic *Lost World* and of movies such as *King Kong* in 1932. Just how, we wonder, would a prehistoric giant dinosaur react if it were turned loose in civilization?

Occasionally a scientist describes an animal or plant as a "living fossil." This label is an oxymoron, because a fossil is something long dead, and no living plant or animal would qualify as a fossil. A living fossil, however, is a living remnant of a long-gone era. This animal or plant is a species that has survived long past the era in which it flourished. In the distant geologic past, usually the family to which this plant or animal belongs was once quite abundant and widespread. Often the survivor has changed little in appearance from its remote ancestors, and the name "living fossil" describes it appropriately. These creatures interest scientists because they provide immense knowledge regarding the appearance of ancient extinct forms. The study of modern survivors has shed light on many an unsolved problem.

Happily this type of creature is not all that rare. Many species are known that have lived long beyond their time with few physical changes from the antediluvian types. The *Sphenodon* of New Zealand, known commonly as the tuatara, is an excellent example of a living fossil. Outwardly it looks like a lizard, but it is unique in many ways. Of many skeletal differences between the tuatara and modern lizards, the most unusual is a third eye, the pineal eye, on top of its head that is a sensory organ. The tuatara is the sole surviving species of a once great order of reptiles scientifically classified as rhynchocephalian (beak head). This reptilian group appeared on earth during the Triassic, the first period of the age of dinosaurs. Fossil records show that this period witnessed the greatest expansion and development of the order and the spread of rhynchocephalians throughout the world. They began to dwindle after the end of the Triassic Period, and their evolutionary line continued along a very limited sphere.

Very little physical change has taken place in the rhynchocephalian for over 150 million years. Except for the living species of tuatara found only in New Zealand, the entire group is now extinct. With the coming of Europeans to New Zealand, even this one species was threatened with the same fate as its ancestors. As civilization quickly spread over the land, the tuatara began to disappear at an alarming rate. It was on the verge of extinction when conservation authorities took the survivors under scrutiny, and they now are rigidly protected.

The modern king crab, *Limulus*, is another survivor from the past. Its earliest ancestors are found in rocks well over 500 million years old from the period geologists call the Ordovician. From that time until the Mesozoic (age of dinosaurs), this group branched out prolifically. Although its fossil history is not thoroughly known, its close resemblance to extinct primeval forms and its being the sole surviving genus of a once great order truly classifies *Limulus* as a living fossil.

What truly captures the attention of the public is the discovery of a living animal whose species has been thought to have been completely extinct for millions of years. Although once a great group of invertebrates, today only one genus, which includes four species, survives. Another living fossil was discovered by a fisherman in 1938. Retrieving his heavily laden fishnets at the mouth of the Chalumna River in South Africa, the man found a strange-looking fish flopping around among the ordinary fish in the net that measured over five feet in length and topped 110 pounds. The weird appearance of this creature caused him to turn it over to Marjorie Latimer, the curator of the museum at East London, South Africa. She also had never seen one like it before, nor was she familiar with any fish of its description. Fascinated, she wrote to James L. B. Smith, the leading fish expert in South Africa.

When Professor Smith finally arrived more than a week later, the now mounted fish was presented to him. His astonishment knew no bounds, for he recognized a creature that was supposed to have vanished completely from the face of the earth some 70 million years ago. This lobe-finned fish was a coelacanth, a type of *Crossopterygii*. The Crossopterygiian order had crawled out of the waters on its lobe fins about 400 million years ago to become the first land-dwelling vertebrate.

Smith had only the muscles, skin, and parts of the skeleton to compare with coelacanths from the fossil record, which are not uncommon. He considered the loss of internal organs catastrophic for his study but reasoned, correctly, that others would be found. Not until December 20, 1952, fourteen years later, did his prediction come true. By that time he was offering a reward of sterling £100 to any fisherman who found another specimen.

Coelacanth number two was caught off the Comoro Islands, between the east coast of Africa and Madagascar. The fish was turned over to the captain of an English vessel, who did his best to preserve it until Professor Smith could arrive on the island 2,000 miles away. The only preservation method available on Comoro was to cut open the fish and salt it thoroughly. This resulted in some damage to the appearance of the fish, but the internal organs remained intact. Smith's patience and tenacity were finally rewarded.

News released to the media with the discovery of the coelacanth identified it as the lobe-finned ancestor of all land-dwelling vertebrates: reptiles, birds, and mammals (including humans). Today's coelacanth is a cousin, many eons removed, of the fish from the fossil record. It has changed little in its outward appearance over these 400 million years. Its body plan is primitive, generalized, and adaptable, so no change has been obvious enough to keep us from recognizing the kinship of this living species to its fossil ancestor.

Since 1952 about 170 specimens of coelacanth have been captured or photographed, all around the volcanic slopes of the Comoro Islands. Their relative scarcity seems to indicate that they may now be on the verge of extinction. Or perhaps they have found their place on the planet: slightly cool, slightly dark, slightly deep (300 to 1,500 feet), and somewhat sterile so competition is minimal. As the volcanic islands are built up and eroded away—give or take another million years—they may discover another place more amenable to their needs or a new way to adapt. The coelacanth did not survive for 400 million years without learning how to change along with an ever-changing earth.

The modern lungfish is yet another example of a living fossil. The six living species even look like something out of the past that should have long been extinct. Their ancestors reached their greatest development about 400 million years ago, about the same time the coelacanths were at a climax. The lungs that they developed are modified swim bladders in most fish. Their somewhat primitive bodies are eel-like with a tail fin (similar to the coelacanth's) that surrounds the rear of the body. The paired fins of some lungfish are broad and capable of supporting the fish; in others they are slender and flexible. Lungfish range from three to six feet in length and weigh up to 100 pounds. They are covered with embedded or large overlapping scales.

Today lungfish live in the swamps, marshes, and rivers of Australia, Africa, and South America. The Australian species live in pools that have a tendency to become stagnant and therefore low in oxygen during the summer, so the lungfish rises to the surface to breathe. The South American and African species live in streams that often dry up during periods of drought. The fish will burrow into the mud, curled up in a sort of cocoon with an opening to the surface where air can seep in. There it will remain during the hot, dry season in a reverse hibernation called estivation. Perhaps these activities offer clues as to how, eons ago, the first fish left the waters to try its luck on land.

Modern living fossil forms include representatives from the plant as well as the animal kingdom. Of all modern flora, the most outstanding living fossil is the ginkgo ("maiden hair") tree, so identified by Asians because of its leaves' resemblance to the maidenhair fern. It has survived down to the present relatively unchanged from ancestors that flourished 250 million years ago. During the Mesozoic Era, when dinosaurs flourished, the ginkgo became very widespread, but toward the end of the era it began to wane.

Few plants have come so close to extinction as the ginkgo has. Prior to the 18th century it was known only from specimens grown in the temple gardens of China and Japan. Since 250 years ago, when it was discovered by European travelers, it has been planted as an ornamental tree in places where it had flourished in earlier geologic time. Able to endure the poisonous fumes of the modern metropolis, this relic of a once great group of conifers is popular along city streets of Europe and North America. In fact the ginkgo appears to thrive wherever it is transplanted and cared for. This is not surprising for a tree whose fossil ancestors have been found in every continent except Antarctica, as well as in New Zealand and Greenland.

Many botanists consider this species to be extinct except when it is kept alive by cultivation. They may be right. If the ginkgo is unable to live in a wild state and depends on whimsical humans for its continued existence, it is truly a dead species. Yet rumors are increasing of stands of native ginkgo in eastern Asia, the mountains of Southeast Asia, and the Chinese interior. Perhaps this remnant of the dinosaur days can still hold its own without the aid of humans. Either way, *Ginkgo biloba* definitely is a living fossil.

Human Counterparts

The use of trickery is not an exclusively human trait; it is also common among other primates. How remarkable to see a limited capacity for communication coupled with a skimpy supply of egotism, insight, and intelligence leading directly to the phenomena of deceit, trickery, and other human frailties. Monkeys and apes provide sterling examples of such behavior. Consider, for instance, the ruse of an ingenious rhesus monkey in New York's Bronx Zoo.

One day this devious monkey simply vanished from its large, rocky primate enclosure. After several days it was recaptured in the nearby park. Keepers carefully checked the fence, the moat, and the rest of the installation for any escape routes. None were found, but the next morning the runaway had disappeared again.

Keepers captured it with no difficulty. This time, after the monkey was returned to its compound, an attendant was assigned to hide nearby and observe the escape artist in action. At daybreak the attendant saw the animal retrieve a banana from a well-concealed hiding place. Apparently the monkey had deliberately saved this fruit as part of an escape plan. With banana in hand, it ran to the broad moat that separated its enclosure from that of its neighbor, a large bull moose. The monkey swung the banana back and forth in short, deliberate sweeps, just as an animal trainer uses a reward of food to persuade an animal to perform.

Sure enough the moose swam over to the monkey. The ever so clever, but apparently water-shy, monkey thrust the banana into the moose's mouth and leaped onto his broad back. Having deposited its "ticket," it rode this ferry to the adjacent enclosure. From there the monkey had no difficulty escaping.

More remarkably human behavior was found in the harem of a baboon in another zoo. An open-air enclosure housed a typical baboon society, in which the strongest male baboon established himself a sultan and denied the other males access to his females. So possessive was the sultan that the slightest flirtation from any of his harem would result in swift, severe punishment.

The sultan could not be everywhere at once. Occasionally, while he was off napping, the females tried to cheat on him even though they were aware of the consequences. One afternoon a female in his harem who had been neglected by the sultan observed that her lord was sound asleep. She quickly took advantage of the situation and began to display her charms quite brazenly. Completely beguiled by her charms, one of the young bachelors began to caress the adulterous female.

Suddenly the sultan reappeared, and the faithless female swung into action with a fantastic act to appear as if she were being harassed. She tore herself away from the young male, gave him a slap on his chest, and fled wailing into the protective arms of the astonished sultan. She appeared to be complaining to the sultan about the aggressive would-be lover, because she kept throwing glances at Casanova, making throaty sounds of fury and drumming on the ground with her forearms. Her trickery worked; the sultan believed her artful pack of lies. Instead of punishing her, he gave the offending bachelor a thorough beating. While the no longer lustful male lay on the ground nursing wounds to both pride and body, the sultan heaped caresses upon his "sorely offended" wife.

Another human parallel among primates is addiction to tobacco. In

1957 chimpanzees in a South African zoo were given lighted cigarettes. Three out of four caught the habit and exhibited many of the peculiarities associated with humans who smoke. They would chain-smoke, lighting one cigarette from the stub of another, or inhale and sit back with arms and legs crossed as they studied the patterns of the exhaled smoke. When the craving was on they would search desperately for an unused cigarette or even a stub; they could recognize a cork tip and light the correct end. What an untapped market for the frantic tobacco industry!

Heavy Nest

A survey of birds' nests of the world clearly shows that most species build a new nest for their eggs each year. Moreover, some will build two or three new homes for their offsprings in a single season.

Not so the bald eagle. This bird uses the same nest year after year. It chooses only one mate for its whole lifetime and with that mate builds a single nest high in a tall tree. Here the pair stays until death does them part. The pair, mostly the male, will keep this single nest constantly in repair by adding more and more fresh nesting material. Constantly built upon, in a few years the nest will become gigantic in size and weigh as much as 2,000 pounds. Scientists have little trouble spotting the large, sturdy nest of bald eagles. They refer to the nest as an aerie.

Offspring, once they leave the nest, never return; they would be looked upon as intruders and dealt with quite harshly. A pair may even kill their old offspring in trying to protect new hatchlings. But if one eagle were to die, the other would seek out another mate. The nest is completely abandoned and a new nest is built and will eventually grow to the same or greater size.

Eagles, by the way, have a unique mating ritual. Seeking a mate, the male will fly around the female, showing off his abilities in specialized flight. Impressed, the female will extend her clawed feet, which the male will lock onto with his clawed feet. Together they will fall, turning and twisting over and over and never relaxing their grip on each other or making use of their wings. When just about ready to hit the ground, they release each other's feet. Using their wings to stop their fall, they fly alongside each other for a while. They appear to have been testing each other's courage and skill. Having passed the test, they are now mated for life.

Planet of the Insects

About 300 million years ago, much of the eastern United States was covered by swamp forests destined to produce this country's enormous coal deposits. That bit of geologic time is known as the age of coal. But plenty of other things were around. These periods (Mississippian and Pennsylvanian) were alive with all kinds of crawling and flying insects. Paleontologists often refer to this time as the age of cockroaches. The quite abundant insects looked very much as they do today, with one minor difference: some of them were about nine inches long. They had their troubles, though, because overhead flew the modern dragonfly's ancestors, which had a wingspan of almost three feet.

Insects have been on earth much longer than indicated by the age of coal. The oldest known insect fossils were found in rocks in central France. These most ancient insect remains, well over 350 million years old, already display well-developed wings for flight. To have evolved such a highly specialized trait as flight, insects must have been around for yet many more millions of years before that.

The diverse ways in which insects evolved and produced many new species is like a fantasy. Presently the earth teems with insects of all size and shapes, but just how great their numbers and how incredibly diversified these small earth creatures are is difficult to conceive.

Scientists have already identified well over a million different species in this century, and about 4,000 new varieties of known species are discovered each year. Entomologists have barely scratched the surface; the final tally is expected to number well over 10 million living species. The meek truly seem to be inheriting the earth. Even more incredible is the intense variation among insect species, ranging from a beetle that thrives in red pepper to another so specialized that it lives only on the tongues of horseflies.

Mars—an Unresortful Planet

The planet Mars is unique in the solar system. Some planets are geologically inert, while others are so active that they obscure their own history; Mars displays geologic features from its birth to recent events. Moreover it is open, through its thin atmosphere, to scrutiny.

Many features of Mars make it a sister planet to Earth. Its day is only 37½ minutes longer than an earth day, and the tilt of its axis, 23.98 degrees, differs less than one-half degree from Earth's 23.5 degrees. The polar ice caps of Mars grow and shrink, and its seasons appear to fluctuate as do those of Earth. Although Mars is only half the size of Earth, its year is 687 earth days, almost twice the length of our year.

The idea of life on Mars has persisted for many centuries. In 1879 the Italian astronomer Giovanni Schiaparelli discovered an extensive network of small, straight lines crisscrossing the planet. He called these *canali*, meaning "channels," but the name was easily corrupted in English to "canals," which implied waterways made by inhabitants of Mars. Astronomers were skeptical, but the idea gained tremendous popular appeal. In the late 1880s Percival Lowell built the Lowell Observatory in Flagstaff, Arizona, to get a better look at Mars and confirmed the "canals." The obvious conclusion was that, in a last-ditch effort to rescue their planet from its desperate water shortage, engineers had constructed a global canal system to transport water from the polar ice caps to equatorial deserts. Dark oases where the canals crossed further confirmed the findings until other astronomers, using Lowell's telescope, discovered the canals were optical illusions.

Belief in Martians dies hard; the *War of the Worlds* scare that accompanied Orson Welles's radio "newscast" in 1938 showed that Earthlings are all too willing to believe in alien life from the red planet. One U.S. vice president's argument in support of the *Mars Observer* space project was that, since scientists can clearly see water canals on Mars and canals are built by rational beings, Mars may possibly support fellow humans.

The *Mariner* probe in 1964 revealed the planet's true face as barren, heavily cratered, and bereft of breathable air or any of the green, purple, or tentacled creatures created in the many science fiction stories about the red planet. Moreover, scientists now know that Mars has no ozone layer to protect it from ultraviolet radiation; therefore its soil has been rendered completely sterile! The space probe *Mars Observer*, destined to orbit Mars in summer 1994, would have answered many

of the questions asked about this planet. Unfortunately the mission was unsuccessful, and the craft is presently lost somewhere in outer space.

Successful missions have shown much of the spectacular landscape of Mars. The planet boasts a "Grand Canyon," Valles Marineris, over

2,500 miles long and more than 5 miles deep that dwarfs the Arizona tourist mecca. Enormous dust storms periodically obscure the planet's features, and shimmering polar caps of frozen carbon dioxide (dry ice) seasonally wax and wane during each Martian year.

Far more spectacular than its ice caps are Mars's breathtaking volcanoes, the largest in the entire solar system. Four gigantic volcanoes are located on a 3,500-mile equatorial blister called the Tharsis Bulge. The largest volcano, Olympus Mons, is 15 miles high and 360 miles across its base, with a caldera 54 miles in diameter. Scientists speculate that these gigantic mountains of fire, now dormant, may be potentially active.

With its towering 78,000-foot volcano (two and one-half times as high as Mount Everest) and its canyon over 20,000 feet deep, the Mars profile is about 100,000 feet. Even with our Marianas Trench 36,198 feet below sea level, Earth's profile is a mere 65,000 feet, just two-thirds that of Mars.

At present Mars cannot support surface water because it would either evaporate or freeze solid. Yet past probes have clearly indicated that water once existed on this planet of extremes. This does suggest that the Martian climate may have once been radically different from the present and may have resembled the climate of Earth. Probes in the 1960s and 1970s photographed branching valleys that looked exactly like dry riverbeds in our desert areas. In addition to the many small branching channels, there are signs that catastrophic floods scoured widespread areas from time to time. Violent eruptions of groundwater appear to have gouged huge channels up to 90 miles wide into the surface. The discharge rates of these raging floods were over a hundred times the rate of flow of the Mississippi River. These great floods seem to have occurred episodically through Mars's early history. The pictures were taken on the three-billion-year-old surface of its southern hemisphere. Almost inescapably, at some time the atmosphere was thicker and the climate was warmer. Perhaps this occurred when the rotational and orbital motions of the planet caused differences in the amount and distribution of solar energy. A few scientists speculate that at such times in its history Mars could have had oceans.

Few doubt that Mars contained much water in its geologic past, but where is it now? Some water has been detected in the atmosphere and some in the polar caps, but most appears to have vanished. Although some scientists suggest that the water might somehow have been blown off the planet, most believe it is still there, locked in ice under the surface or trapped in clay minerals.

However warm and humid the climate might have been in the past, it is far from that at present. By Earth's standards Mars is a forbidding place. Summer temperatures at noon over the equator could reach a pleasant 68 degrees Fahrenheit, but because the Martian atmosphere (consisting mostly of carbon dioxide) is so thin, the nighttime temperatures can plunge several hundred degrees. During a Mars winter the temperature at polar regions will hover at minus 220 degrees Fahrenheit. There will be no sunbathing there!

"Mighty Mouse"

In 1968 a 12-foot black mamba went on strike at the Transvaal Serpentarium. Resisting attempts of inept attendants to "milk" it for its venom, it bit one of the workers on the wrist. The worker was hurried off to the hospital, where quick medical treatment with mamba serum barely managed to save his life.

Back in its glass pen, the mamba reared up and struck at the glass every time anyone came within vibrating distance. The director of the serpentarium was concerned that the mamba would damage its fangs and become useless for milking. After all, he has paid the equivalent of $1.50 per foot for this particularly deadly reptile, and any damage to its fangs would mean sacrificing his investment. To distract the nettled mamba, the director ordered a chunky white mouse thrown into the snake's pen as a snack.

Normally a frightened mouse would capture the attention of a hungry mamba, even one with an aching mouth. But the stratagem got off to a bad start. The mouse, far from being terrified, ran up to the rearing snake and bit it firmly on the side. The rodent then retired to a neutral corner, completely ignoring the snake, and busied itself feeding on scraps in the pen and cleaning its whiskers. Two days later it was still finding scraps and grooming itself, but it was alone in the pen. The deadly black mamba had died of infection from the mouse bite!

The Fish That Almost Changed History

Almost everyone is familiar with the story of Pocahontas, the vivacious, audacious, Native American teenager who rescued Captain John Smith when his head already lay on the executioner's block. Placing her head between his head and the warriors' clubs poised to bash out his brains, she brought about a last-minute reprieve. Pocahontas's father, Chief Powhatan, moved by her act of courage, spared the captain's life and arranged a truce between his tribe and the Jamestown colonists.

This event-turned-legend has been the basic ingredient for many films and literary works, although the details of the actual incident may have differed from those described above. However, this romantic bit of early American history would not have occurred if an earlier alarming incident in Captain Smith's life had turned out as unpleasantly as it might have. The occasion was his unfortunate encounter with a dangerous fish known as the stingray.

Stingrays are common in shallow tropical to temperate, fresh or marine waters throughout the world. They have been known and feared for many centuries. Pliny the Elder, in the first century A.D., wrote that "its spine wold pierce armour like an arrow, and, driven into its root, wold cause a tree to wither. To the strength of iron it adds the venom of poison."

Although Pliny's description is not entirely accurate (unless stingrays have tamed down considerably since his day), the stingray accounts for most of the injuries caused by venomous fish, about 750 per year in the United States alone. Despite Pliny's claim that the stingray "lurks in ambush and pierces fish as they pass," it is nonaggressive and spends most of its time half-buried in sand, where it finds such favorite foods as mollusks, worms, and crustaceans. Because the stingray lives in places frequented by bathers, beachcombers, and fishermen, victims are usually wounded when they inadvertently step on a buried fish. Obviously not amused at being aroused or stepped on, the stingray immediately lashes its whiplike tail in all directions, randomly piercing anything soft and accessible—such as a human leg.

The flat, disklike body of the stingray comes in three shapes: round, kite, or diamond. Approximately as wide as they are long, stingrays range in size from 12 inches to 14 feet, and weigh from 1.5 to 750 pounds, depending on which of the 100-plus species a person may encounter. The slender tail is at least as long as the rest of its body. On the upper side of the tail is the stingray's weapon: a sharp, stiff poison spine, or dart. A venomous secretion flows in grooves to the underside of the spine, which is fringed with thin barbs that point backward.

When the stingray stabs, it not only injects poison but also cuts and tears the flesh. Even a tiny puncture from a stingray's spine has been known to make an adult victim lose consciousness. The effect of the poison is immediate, and inflammation spreads around the wound almost as soon as the spine has penetrated. The venom produced by the stingray affects the cardiovascular system and causes a loss of blood pressure along with an increase in heart rate.

Although rare, fatalities do occur. One such incident happened off the Bahamas. A scuba diver was foolhardy enough to pick up a two-foot stingray and bring it close to his body. The tail immediately began a frenzied lashing; piercing the diver's chest, the spine entered his heart and he died instantly. This case delivers a message quite appropriate for most objects of nature: look at it—don't touch it and don't pick it up!

In 1601, six years before Pocahontas saved him from Native American war clubs, Captain John Smith almost suffered an agonizing, unhistoric death when he was stabbed by a stingray. His ship had just run aground on a shoal in Chesapeake Bay. While the crew set things to right, the captain amused himself by wading in the shallow waters and stabbing fish with his sword. A small stingray drove its spine into the captain's wrist as he tried to retrieve his sword. Instantly ill he barely made it to shore, where he collapsed.

In his *General Historie of Virginia*, Captain Smith described in third person his harrowing adventure. The captured stingray had

> *a most poisoned sting . . . which she stucke into the wrist of his arme near an inch and a half; no blood nor wound was seene, but a little blewe spot, but the torment was instantly so extreeme, that in foure houres had so swollen his hand, arme and shulder (that the crew) prepared his grave on an island near the mouth of the river. . . . But ere night his tormenting paine was so well aswaged that he eate of the fishe to his supper.*

To his companions this was indeed miraculous. He had been in such agony they had even considered putting him out of his misery.

American history was back on track, and Captain John Smith stayed around to earn his paragraph in the chronicles of our country.

Bear Hunt, Stone Age Style

In Drachenhohle Cave, Switzerland, bears were frequently hunted and killed by Neanderthals. The cave contained at one point a very narrow corridor, through which the bears could not avoid passing and where they were really quite helpless because of the restricted space.

The hunter waited at the end of the corridor, hidden behind a boulder. As the bear emerged the hunter would bash in its head with a club of some sort. The fact that so many of the bear skulls show the damage on the left side indicates the concealing boulder was on that side of the corridor's exit point. One skull still has a 60,000-year-old Neanderthal stone ax point embedded in it.

The Cro-Magnons did one better. In Sloup Cave, Moravia, a bear skull was found with a nicely healed injury to the crown caused by a hunter well over 20,000 years ago. Nearby lay the broken spear point that had pierced the bear's skull. The point had remained in the skull until the bear's death years later, when postmortem decay caused it to fall out.

One can easily envision what happened in this Paleolithic event. The hunter, defending himself from the attack of the huge bear, possibly tried to drive the beast from the cave. The hunter lost because even though the spear point went deep into the bear's skull it did not stop the huge animal's charge, and the caveman quickly joined his ancestors. The victorious bear had to go through the remaining years of life with a perennial bear of a headache.

CHAPTER NINE

Gold Fever

"Thar's gold in them thar hills."

The Valley of the Kings is a rocky wasteland on the western bank of the Nile River, across from the ancient Egyptian capital of Thebes. In this valley most of the great Egyptian kings and nobles were buried in utmost secrecy to prevent graverobbing. The elaborate, clandestine schemes were to no avail; by the end of the 20th Dynasty almost all of the royal tombs had been opened illegally by professional thieves (probably with the help of the high priests). Plundering by tomb robbers was so thorough that most royal treasures had vanished long before early archaeologists arrived on the scene.

As usual there was an exception. In 1922, when Howard Carter peered through a slit he had made in a door, he was about to unearth the greatest treasure trove in all history. He had discovered the almost undisturbed and long-forgotten tomb of Tutankhamen, the boy king who died in 1339 B.C. at age 18. Although Tutankhamen was a lesser king, the total riches found in his burial chamber are mind-boggling. The mummy was encased in a nest of three coffins, the inner coffin being the most spectacular. Just over six feet long, it weighed 2,448 pounds and was composed of solid gold! As a work of art the coffin is priceless, but the value of the gold of which it is molded would be, at current fluctuating prices, over $10 million. Despite the tenacious permanence of gold, most of the royal gold of Egypt has disappeared over the millennia. But we can barely comprehend the incredible wealth that must have been buried with great pharaohs such as Ramses II. Almost all the gold was in the hands of the nobility. Because all precious metal immediately became the property of the ruler, gold has rarely been found with the remains of ordinary Egyptian citizens. Gold was the symbol of worth for royalty. It showed their kinship to the sun. Being encased in a body of gold after death guaranteed immortality.

Gold was a scarce commodity in Egypt. Except for the gold acquired from conquered nations, the mines of Nubia and the Arabian peninsula were the pharaohs' only sources of the ore. Gold is still a very rare element, with many uses; the demand always exceeds supply. The earth's crust contains only 3.5 parts per billion of gold. Distribution is worldwide, with important deposits in North America, South Africa, and Russia. Despite the gold found in the archaeological remains of many early civilizations, more gold was mined during the 19th century than during all the preceding 5,000 years, and over two-thirds of the world's known gold was mined in the 20th century. The total harvest of gold throughout history is about 100,000 tons. Because gold is durable and quite recyclable, most of the gold that has been discovered still exists. Your gold tooth or bracelet may have been a prized amulet of a Stone Age person, a headdress for an Egyptian queen, or an Inca statue.

Today ownership of gold is not restricted to royalty, so the lure of the most noble metal affects multitudes of people. In the American southwest, one of the common hobbies among outdoor enthusiasts is searching and panning for gold. Mountain and desert areas, with their sparse ground cover, are the most vulnerable to weekend treasure seekers. The effect on the terrain is disastrous; ground is torn apart by people searching, searching, and still searching. In the quest for gold, much of the California desert area has been claimed by individuals for potential mineral exploration. So a widespread search continues, the natural environment suffers, and the endangered and threatened species list gets longer.

In 1989 a hearing was conducted in Barstow, California, for officials to learn the pros and cons of the Desert Protection Act. Miners and prospectors, along with hunters, bikers, ranchers, and others who did not want their desert operations curtailed by government regulations, appeared in droves to register their opposition. One professional witness for preservation was a registered geologist who had been a mineral consultant for nearly 30 years. He explained that he had conducted mineral surveys on nearly 200 properties, usually for a person who had found a few flakes of gold on his or her property and was convinced that this was a true bonanza. Out of nearly 200 claims, only 4 had been economically sound. Because all the easy-to-reach gold had already been taken, it would cost far more to extract any remaining gold than the market value of the mineral. As the geologist pointed out to the audience, "Two nuggets do not make a mine!" His remarks failed to convince the rockhounds and prospectors, for whom the lure of gold was so intense that they were willing to sacrifice almost anything for it. Few would ever experience the thrill of finding the real thing.

Several years earlier a government geologist was conducting a geologic survey just outside Death Valley National Monument. He picked up a strange-looking rock, sort of green-black and out of place in the area. Noting how heavy it was, he tossed it into his pack for a more thorough examination later.

In the laboratory, the geologist picked up the two-inch piece of rock and turned it over for the first time. The opposite face was solid, smooth gold; his startled face stared back at him. As others have done, he convinced himself that "one nugget might actually make a bonanza." Unfortunately he was in Tucson, Arizona, a long way from the western slope of Death Valley.

A few years later, the geologist became a professor of geology at a California university. With him went the piece of gold with the mirrorlike surface, which now served as a paperweight on his office desk. On several occasions he returned to the place where he had picked up the gold. He traversed a large area, but no further gold was anywhere to be found. Apparently this chunk of gold had been carried there and dropped years before.

Knowing Death Valley was the area of the famed lost Breyfogle Mine, the geologist often considered the possibility that the gold was part of the treasure dropped by the unfortunate Charles Breyfogle as he tried to find his way back to civilization. People can easily become confused in the desert, even when they're used to the environment. Volumes have been written about lost mines and buried treasures. The deserts are pockmarked with the endless searches, and the results are very predictable: the discoverer spends years trying unsuccessfully to relocate the original find—with an almost 100 percent failure rate. That Charles Breyfogle was unable to find his way back to his treasure is therefore no surprise.

One spring day in 1862 the horse of a Pony Express rider racing across Nevada stumbled to his knees. While recovering his balance, the horse kicked loose a chunk of rock that caught the rider's eye. The man took it to Virginia City, where it was pronounced to be silver ore of extraordinary richness. Suddenly a great silver rush began around the present town of Austin, Nevada. When the news reached Los Angeles, three down-on-their-luck men, with neither funds nor means of traveling, determined to head for the Austin area and sure wealth. They were remembered only as McLeod, O'Bannion, and Charles Breyfogle.

Around June 1, 1863, the trio set out on foot for the silver fields that lay 400 miles to the northeast over forbidding, desolate mountains and deserts. The men carried scant provisions as they crossed the Mohave Desert, skirted the spurs of the Argus Range, crept across the glittering wasteland known as the Panamint Valley, and finally began the ascent of the awesome Panamint Mountains. From the top they could see to the east the unearthly basin known as Death Valley.

On the eastern slope of the Panamints they found water and decided to spend the night. The search for a place that was not too rough and sloping was almost impossible in this tricky mountainous terrain, but they finally settled in a small plot not far from the water hole. McLeod and O'Bannion laid out their pallet together, and Breyfogle found a bedding place about 200 yards down the slope. All prepared for a restful night's sleep by removing only their shoes.

During the night, Breyfogle awoke to moans, screams, shouting, and the sounds of clubs smashing heads. He realized that his partners were being attacked and doubtless killed by Indians, so he sprang from his blanket, grabbed his shoes, and, abandoning his other supplies, fled barefooted to the valley below. Wild with fear, he ran over rocks and thorns through the desert night, finally limping at daybreak to the bottom of the mountain and the floor of Death Valley.

By afternoon Breyfogle found a geyserlike hole containing alkali water. He drank it, the first water he had tasted since the previous evening, and it made him deathly sick. But it was moist, and he recovered, so it was better than nothing. Since his feet were too bruised and swollen for his shoes to fit, he filled them with the tainted water and went on. Still fearful of the Indians he continued walking toward the northeast for two days, with only the alkali water for sustenance, even though it made him sick and did little to quench his thirst. Near the mountains he saw a green spot that might mark a spring; he estimated it to be about three miles away. The shortest route was through an extensive sand dune area now believed to have been the dunes near

Stovepipe Wells. About halfway to the green spot, he noticed, where wind had uncovered a layer of bedrock, a layer of soft greenish rock shot through with smoky quartz. Free gold showed plainly through the rock.

Breyfogle almost forgot his plight as he turned over a rock and stared at its mirrorlike gold surface. Unmindful of his fear of the Indians, his exhaustion from the tortures he had endured, or his craving for fresh cool water, he marveled at the promise of wealth that exceeded his wildest dreams. He gathered several of the richest pieces and wrapped them in his bandanna. Then he staggered on toward the green spot, which proved to be a low, bushy mesquite tree full of green beans. He ate ravenously of them, but they did not satisfy his intense need for fresh water, and he collapsed.

The next few days were an absolute blank to him. At the clear, fresh water of Baxter Springs, fully 250 miles from where he had emerged from Death Valley, Breyfogle became conscious. He remained there for two days, drinking freshwater and eating whatever edible vegetables he could find. When his strength returned he set out once again toward the silver strike area.

A rancher named Wilson ran across Breyfogle. For many years afterward his description of the human wreck standing before him was a fireside story familiar all over eastern California and Nevada. The most outstanding part of the description was the bandanna Breyfogle had stuffed in a shoe. It was tied around the richest gold ore Wilson had ever seen.

Breyfogle remained at the ranch until he regained his health. Afterward Wilson got him a job in a mine near Austin, Nevada. Months later Breyfogle, the manager of the mine, and six other men set out for Death Valley to find the treasure. They reached the Funeral Range and were about to cross into Death Valley, when they met hostile Indians. The treasure seekers wisely returned to Austin.

Almost a year later Breyfogle and his group set out again for Death Valley. He had no trouble getting back to the area where he had feasted on the mesquite tree. But many similar trees grew in the area, and so he wavered. He knew the gold had to be close by; the men frantically searched in every direction and nerves were frayed. They began to jeer Breyfogle, and curse him for having led them on a wild gold chase. Finally the men packed up and returned to Austin.

Breyfogle made several more attempts to find his hidden mother lode; party after party searched but to no avail. He died penniless in 1870, but the desert rats have continued ever since to search out his gold.

George Hearst, father of publisher William Randolph Hearst, kept men in the field searching for the lost mining site for years. Others followed, but the treasure has never been found.

An incident happened just a few years after Breyfogle's death that could possibly explain why the gold has been so elusive. Near Yuma, Arizona, a shepherd girl found herself caught in a violent sandstorm. To protect herself she lay flat on the ground while the howling wind raged. She felt the sands being carried away around and under her until the surface she was lying on felt hard and rocky. When the storm abated the girl saw that the little rocks around her were solid gold nuggets. When she returned to her employer's ranch and showed the nuggets, horses were saddled in record time.

The ranchers never got to the place where the treasure lay, because another sandstorm hit. When the storm subsided the entire topography of the sand dunes had changed. The bed of nuggets now lay under one of the many sand dunes, and the gold was never rediscovered. Doubtless, while Breyfogle waited in Austin, cloudbursts and sandstorms also covered his rocks of gold.

What the desert sands cover one day they will uncover again the next. Someday a visitor to Death Valley, seeing a golden glint among the dunes around Stovepipe Wells, may rediscover Breyfogle's treasure. Minus, of course, one mirror-surfaced nugget that the geologist continues to use as a paperweight. Moreover, now that Death Valley has become a national park, Breyfogle's gold actually belongs to all of us.

"How Doth the Little Crocodile"

The crocodiles and their close relatives—alligators, caimans, and gavials—are the sole survivors of the great group of reptiles, the Archosauria, which included the awe-inspiring dinosaurs. Indeed the crocodile, which continues to lurk in rivers and swamplands of the tropics, is a living fossil of the Mesozoic (the dinosaur era) of more than 100 million years ago. The first true crocodiles, such as Phobosuchus, the "horror crocodile," were fearsome giants up to 60 feet long, with heads measuring over six feet. Lurking in primeval swamps and estuaries, they likely waxed fat on baby dinosaurs (and an occasional parent).

The modern crocodile, smaller than its ancestors, has also been a terrifying figure. The ancient Egyptians, obsessed with death, venerated the crocodile as a death symbol. Paintings in tombs show the priests feeding special cakes and honey wine to the crocodile, which they regarded as the incarnation of the water god Sebek. The Egyptians embalmed hundreds of crocodiles, decorating their scales with gold rings before laying them to rest in sacred tombs.

The crocodile's domain includes central and southern Africa, the warmer parts of Asia, tropical Pacific islands and northern Australia. Crocodiles and alligators are easily confused, but the two species have distinct profiles. With its jaws closed the teeth of the alligator cannot be seen; crocodile teeth are usually on display, imparting the famous and characteristic crocodile grin. The fourth tooth on each side of the lower jaw, almost tusklike, fits into a dent in the snout and is visible at all times. In general the crocodile has a longer, narrower snout than that of the alligator, which is broad and blunt. Most crocodiles measure from 10 to 20 feet and weigh up to 2,000 pounds. The Pacific saltwater crocodile has been reported (although not confirmed) at over 28 feet in length and weighing 4,400 pounds.

The crocodile, the largest of reptiles, is the most feared beast in the modern world. Many experts believe the crocodile kills more human beings than do tigers, lions, leopards, and snakes put together. Each year in Africa alone, an estimated thousand people are pulled down by crocodiles. When a human-eating cat infests a region, people band together for safety and take every possible precaution, but somehow a fatal carelessness is often manifested in the case of man-eating crocodiles that haunt local waters. Crocodiles will not deliberately seek out human prey, but they do respond to the opportunity. This is confirmed by the fact that most victims are women bathing, washing clothes, or drawing water and children splashing in the shallows.

A crocodile that catches a large animal keeps the victim off balance with a savage sideways yank of its head and pulls the animal into the water. Drowning soon stops the victim's struggles, and the crocodile may roll over and over so that the victim is dismembered. Impressive as the great jaws may be, a crocodile's teeth are not made for chewing. It can therefore eat only something it can swallow whole. Small dogs are a favorite tidbit. If the victim is large, it is towed away to rot and thus become soft enough to tear apart easily. The crocodile will often take prey to a tunnel-like den with a below-water entrance. The den slants upward underground above the waterline, and has an air vent in the roof.

In East Africa a boy about 12 years old was playing at the river's edge when he was seized by a crocodile and pulled into the water as his friends watched helplessly. By enormous luck the reptile's den was only a few yards away. The victim regained consciousness to find himself inside a dim cavern full of decaying carcasses, the crocodile lying alongside him. The boy didn't move a muscle! When the reptile went back into the water, the boy seized the opportunity to enlarge the airhole above him and escape.

The boy's family, sure he was a ghost come back to haunt them, would not let him into the house. He ran to the hut of the chief who, after overcoming his own terror at the sight of the boy, realized the youth had miraculously escaped the crocodile. The chief then went to the family hut and ordered the inmates to open the door to their kin. That night a special feast was held to celebrate a miracle; the main dish was roast crocodile.

Most widespread of the 13 species of crocodile is the Nile crocodile of Africa and Madagascar. It breeds when 5–10 years old by which time it is 7–10 feet long. The male stakes out a territory along the river bank and will defend this bit of land from all intruders. The fights can sometimes be deadly.

To mate, the male approaches a female crocodile and displays himself to her by thrashing the water with his snout and tail. If she approves, the two reptiles swim alongside each other in circles with the male on the outside trying to get near her so he can put a forelimb over her body and mate.

The Nile crocodile digs a pit two feet deep near water and covers the eggs with sand, usually in the shade where the female can guard her brood and keep herself cool. She then stays by the nest during the incubation period defending the eggs against all enemies—including other crocodiles, although they sometimes nest in colonies only a few yards apart. The mother has to be on her guard at all times. Many animals wait for a chance to eat the eggs or the baby crocodiles. Their most aggressive enemy is the monitor lizard, which is occasionally bold enough to dig underneath the mother crocodile as she lies over her nest. Once a male monitor was seen to decoy a mother away from the nest while the female stole the eggs.

In due course the mother hears a hiccuping signal from the hatching young and opens the nest. The babies have an egg tooth on the tip of their snouts with which they break through the tough shell. The female's instinct to dig away the sand covering the nest is so strong that when scientists tested it by building a fence around a nest, the frantic mother tore it to pieces.

Up to 90 eggs are laid during the dry season and hatch three to four months later during the rainy season, when the babies have plenty of insects on which to feed. The newborns, about 10 inches long, are aggressive from the first moment of life, snapping at anything near them. With unerring instinct they head for the nearest water. When necessary for their safety, the mother will carry them to the river's edge in her mouth. Few will live beyond infancy; many predators, including storks, cranes, turtles, predatory fish, and even other crocodiles, find these wiggling infants very tasty. Some crocodile experts estimate that no more than 1 percent of the eggs laid will become mature crocodiles.

Despite its fearsome reputation the crocodile is completely docile at times. The Nile species will often lie on the riverbank with its gigantic jaws wide open. The Egyptian plover, a small bird about the size of a sparrow, flies or walks into the open mouth. Standing on the crocodile's tongue it performs the useful task of picking debris from the crocodile's teeth. This symbiotic relationship has been thoroughly studied by scientists. In fact, whenever these birds are at hand crocodiles will open their jaws deliberately, an evident invitation. Observers have never known the crocodile to close its jaws on the little bird walking into its mouth.

Reptiles are classified as cold-blooded because they can't maintain body temperatures as can mammals and birds. A crocodile can't shiver to keep warm or sweat to keep cool. Its body temperature is usually within a few degrees of its surroundings. So it follows a daily routine to avoid extremes of temperature. Crocodiles come out of the water at sunrise to lie on the banks and bask in the sun. When they have warmed their bodies they move either into the shade or back into the water to avoid the full strength of the midday sun. In late afternoon they bask again, returning to the water by nightfall. Since water holds heat better than air does, the crocodile remains mostly underwater throughout the night.

Basking crocodiles have to worry about cannibalism among the species. So they bask in groups of equal-size crocodiles. Smaller ones take care to keep well away from larger ones. They instinctively know that basking with larger crocodiles will almost certainly be the last thing they ever do.

After World War II the market for crocodile leather products such as shoes, handbags, and luggage was tremendous. This spurred widespread harvesting of the crocodile. Governments have put restrictions on hunting and exporting, but the crocodile is dwindling worldwide. Since many species are valued for their flesh and native people depend on them for essential food, crocodile farms have sprung up to keep pressure off the wild populations. Many lakes, swamps, and rivers now almost bare of crocodiles may look forward to the excitement of their return.

War Stories

Knowledgeable, experienced General William Sherman defined war succinctly—"War is at best barbarism"—and unequivocally—"War is hell." Those who have lived through it will agree that war is about as bad as things ever get. But occasionally events that happen during a war can make it even a little worse.

Tempest, not Temper, Control

In the year 483 B.C. Xerxes the Great, king of Persia, assembled an enormous army and armada to fulfill his father's thwarted plan to conquer Greece. By 480 B.C. his forces had arrived at the strait of Dardenelles (Hellespont). To carry his armies rapidly across the narrow waterway (less than a mile in some places) he ordered his engineers to erect a pontoon bridge of boats. At first the crossing went well, but an intense storm came up and wrecked the bridge. That the defiant storm should join forces with the Greeks was unacceptable to the uncompromising, imperious Xerxes. Therefore he ordered his men to give the Hellespont 300 lashes, and to throw a symbolic pair of shackles into the waves. This didn't seem to intimidate the raging waters, so he had the water branded with hot irons. In his arrogant perception, the hissing sounds were the water's cries of pain. The engineers who had built the bridge were beheaded for their faulty workmanship.

The Hellespont, obviously no match for Xerxes, calmed down. Newly appointed engineers rebuilt the boat-pontoon bridge and kept their heads.

Xerxes' invasion was initially a success, but he suffered a crushing defeat at Salamis later that year and was forced to implement a strategic withdrawal. The river gods were vengeful and whipped the Hellespont into an even greater storm. Knowing his days would be short in number should he remain in Greek territory, he made a long, tempestuous crossing. Safely at home in Persia and wallowing in a life of pleasure such as great kings deserve, Xerxes was murdered by palace guards.

Enemy Secret Weapon

A veteran of World War II recalls the time that almost the entire platoon came down with food poisoning. All were so miserable that they agreed later they would have considered an enemy attack an act of mercy. As they were behind enemy lines, any attack would have been a total disaster. There was no field operation that day; the only action was the medic making rounds as he administered morphine, the available painkiller. Most soldiers recovered by the next morning, but the sore stomachs hung on for several days.

This kind of wartime double whammy was a big reality during the Spanish-American War, when one of America's greatest food scandals took place. Over one thousand U.S. soldiers died from eating tainted canned beef, a most miserable way to die. In the final reckoning, more casualties resulted from "embalmed" beef, as it was called, than from enemy action. Only 279 U.S. soldiers died in combat from enemy gunfire.

The tainted beef caused quite a scandal back in the United States, but the results were favorable. This tragic incident led to the passage of the Pure Food and Drug Act in 1906, requiring government inspection of all packaged food.

Bees as a Weapon of War

The use of bees in human conflict is probably as old as civilization itself. Records show a number of incidents in which bees were used as weapons in warfare. For example, when Henry I was besieged by the Duke of Lorraine, the garrison's commanding general ordered beehives thrown among the horses of the attacking forces. The bees caused such confusion among horses and soldiers that the siege was quickly lifted.

Richard Coeur de Lion (the Lionhearted) reduced the citadel of Acre by a similar means. He ordered beehives to be hurled among the defending garrison. To escape the angry bees, the defending soldiers took refuge in their cellars. Thus Richard's men easily battered their way into the defenseless citadel—and were almost driven back by their former bee allies.

Bees have put soldiers to flight in more modern times. In India a cavalry soldier accidentally put his lance through a beehive. The angry insects swarmed out of the hive and attacked the nearest group of men, a company of kilted Highlanders whose bare knees were easy prey. The soldiers broke and ran, leaving the Indian troops to celebrate victory.

During World War I, German soldiers employed bees successfully as a weapon against the British on several occasions. They gave the bees a reprieve during World War II.

Angry Fire Gods

Accounts of the last days of Pompeii have been recorded over and over again. Recall that Mount Vesuvius destroyed the famous city in A.D. 79. The volcano then was less active for over 1,500 years, with moderately violent eruptions occurring about once per century. After an eruption in 1139 no activity occurred for about 500 years, so Vesuvius was thought to be extinct.

For centuries people settled the surrounding countryside and cultivated the fertile soil of the volcanic slope and plain. The land was alive with activity. Unfortunately, so was the volcano; on December 16, 1631, it erupted violently. This eruption may have been even stronger than the historic activity of A.D. 79 and was definitely more destructive. Over 18,000 people were killed, most of them farmers who lived in the shadow of Vesuvius. In a matter of minutes a cloud of incandescent gas and ash exploded down the side of the volcano at hurricane speed, engulfing neighboring farms and towns.

After that great convulsion Vesuvius was never totally inactive; it erupted several times but never so violently. People grew accustomed to its controlled fury and gradually moved back into the vicinity, although their sense of history encouraged them to maintain a respectful distance from the volcano.

Vesuvius decided to let go again on March 18, 1944, at 4:30 P.M. Italy was already having its share of headaches, especially since the Allies had gained control of the area. But this last great eruption temporarily halted the war. Everyone retreated to volcanically neutral territory, and no further fighting occurred until Vesuvius's fury was spent. That was several days later, after the explosive phase had reburied the ruins of Pompeii under almost a foot of ash. No army dared to compete with the ferocity of the fire god!

The Time Justice Was Just

Even in total war, justice seems to prevail at times. So it was with Baron Von Schlavrendorff in early 1945. The man had been brought to trial before a Nazi court and charged with an attempt on Hitler's life, a charge of which he was probably guilty. Had he been successful, the war would have ended sooner, and thousands of lives would have been saved. In a very short trial, the baron was found guilty of high treason and sentenced to death. The court ordered that the execution by firing squad take place at once.

As the baron was being led out of the courtroom, a surprise Allied air raid began. Before anyone could run to shelter, a bomb demolished the court building. All connected with the trial were killed instantly by the explosion. All, that is, but one man.

Incredibly the blast that killed everyone else in the courtroom spared the life of the condemned baron. He was knocked almost unconscious but realized, when his head cleared, no one was around to hold him. All his captors lay dead in the rubble of the courtroom. The baron fled the shattered building and successfully found his way to freedom.

Adaptation for Survival

Many upland regions near the equator lack the change of seasons most people are accustomed to. In fact, in certain areas of the Peruvian and Bolivian Andes the climate is forever springtime. Temperatures are less extreme, rainfall is distributed more evenly throughout the year, and the soil is more productive than that of terrain near sea level. Here fruit trees bloom throughout the year, and many field crops may be harvested several times annually. Mountain habitats also are free from tropical diseases, since high altitudes are unsatisfactory to the needs of mosquitoes, tsetse flies, and other scourges of the tropics.

The environment in these areas was friendly enough to have fostered the great Inca civilization. In the course of its 4,000-year development, millions of people occupied the high valleys of the central Andes. Many established some of the highest permanent habitations on earth.

At extreme altitudes, however, the advantages of equatorial mountain living vanish. The air is too thin and cold to be considered friendly. For the ordinary person, the climate in the high Andes is so unfavorable that the problem of survival is almost unsolvable. The dilemma: how does one get enough air up there to breathe?

Most of the human population of the earth lives near the bottom of a 10-mile-deep ocean of air. This air has weight and is easily compressed; therefore it becomes denser as it gets deeper. At sea level humans are adapted to a density (atmospheric pressure) of 15 pounds per square inch. Sea-level humans' lungs are so conditioned that when one breathes in a gulp of air the 15-pound pressure forces a significant supply of oxygen through the thin linings of the lungs, giving the breather what is needed to function properly.

At higher elevations air pressure is reduced, and at 10,000 feet it is only 10 pounds per square inch. This is not quite enough to push an

adequate supply of oxygen through the linings of the lungs. As a result blood carries up to 15 percent less than its normal load of oxygen, a short supply that may cause headaches, fatigue, and shortness of breath, commonly called "mountain sickness." At 18,000 feet air pressure is only half that at sea level, and few people escape more pronounced symptoms. If the human body is unable to overcome the shortage of oxygen, it will sicken and die.

Yet residents of the high Andes not only survive at such altitudes but spend their entire lives there, working and playing as normally as do most people at sea level. Clearly these people are benefiting from the adaptions of generations of their predecessors to an extreme environment. Some changes they have made to overcome the altitude problems are dramatic and permanent. They have become physically different from the rest of us, mainly as a result of changes in their respiratory and blood-circulating systems.

A visitor to the high Andes will immediately notice that the natives of these mountain slopes are short and stocky and have tremendous, barrel-shaped chests. They have lungs to match; the small pockets (alveoli) that line the lungs and give added capacity are always wide open in these high-altitude dwellers. This feature allows the greatest possible amount of blood to flow through the delicate lung tissues, thereby picking up all available oxygen that is breathed in.

The natives of the high Andes have about 20 percent more blood than a lowlander. An Andean mountaineer and a sea-level dweller, both weighing 130 pounds, would compare as follows: although the Andean is shorter in stature, his total blood content is about six quarts, more than half red blood cells. The lowlander has only five quarts, red cells making up about 40 percent. The Andean's extra red cells hold extra hemoglobin that can catch and absorb oxygen. People living at altitudes of 15,000 feet or higher have an almost 60 percent increase in this vital hemoglobin. Moreover, each of their red blood cells is larger than the cells of lowland people; this larger size gives a greater surface for the absorption of oxygen.

Since the mountaineer's blood is so rich in red cells, it is of course thicker than the lowlander's. The heart of the Andean is about a fifth larger because it must pump harder at these altitudes, but the beat is much slower. The blood doesn't have to be pumped very far as it circulates through the highlander's squat, compact body.

The ends of the Andean's broad, stubby fingers and toes contain an unusually large number of direct passages between arteries and small veins. This hastens circulation, enabling the mountain person to go about with hands and feet uncovered at times when lowlanders would suffer severe frostbite. As can ducks and penguins, these people walk barefoot in the snow without discomfort.

In 1981 a team of scientists from the United States was working on a newly discovered Inca ruin in the high Andes. The expedition cook, who had just gotten married, planned not to report for work the next day but promised to send his brother to fill in for him during his brief honeymoon. The scientists were disgruntled the next morning because breakfast time came and the replacement had not arrived. Since it had snowed during the night everyone was cold, hungry, and uncomfortable. A cheer went up as the new cook arrived on the run with an apologetic grin on his face; he had overslept. Then scientists realized that the man had plowed barefooted through the deep snow. Because he was late, he didn't take the time to put on shoes and carried them instead. Immediately the scientists, laced in warm boots, felt more comfortable.

Insect Jewelry

People use up much free time looking for the latest in decorative jewelry. Both women and men spend billions of dollars for decorative items. U.S. women in particular look for hair decorations that are exotic and different.

Some of the poorer people of the world find ways of looking attractive that are far less expensive than objects of adornment sold in U.S. department stores. These people turn to nature for their decorative items. For example, some of the young women living in Costa Rica will scour the forests and bushes at night for the most naturally luminous insects. The women secure these creatures with very tiny chains or cords, allowing them some leeway of movement, and fasten them to hair or clothing in the evening. As the insects crawl about they flash their varicolored lights, creating a most striking effect, one almost impossible to create inexpensively in a department store. The female adorned with nature's beauty is most attractive and noticeable.

The young women of Malaysia manage to go one better than the Costa Ricans because the natural ornaments they use can be seen also in the daytime hours. In this country some of the butterfly fauna have extremely beautiful wings and bodies. Many of these colorful species are used as ornaments. They are gently tied to the woman's hair, with a slight leeway, by tiny chains or by very thin cord or thread. As they periodically attempt to fly away or just slightly beat their wings, they create a most unusual display of nature's beauty. Male observers are immediately attracted to the bearer of such natural charms. This kind of beauty one cannot find in a bottle or jar.

Written in the Rocks of Time

Scientists have been able to learn much about the physical characteristics of fossil animals from their bones and other hard body parts. Lifestyle, however, is often more evident from footprints. Fossil footprints record events that occurred thousands, even millions of years ago.

In Australia a series of recently discovered dinosaur tracks tell of a prehistoric stampede. The tracks show a group of dinosaurs that first were milling around, probably grazing. Abruptly the dinosaurs turned and began to move in a single direction. They were obviously running, for the tracks become deeper and farther apart. We know their lives

were at stake. Superimposed on their tracks are those of a large carnivorous dinosaur. Whether the carnivore's attack was successful will never be known, since the tracks showing the final scene of the hunt have eroded. This event occurred over 120 million years ago!

On a limestone cliff outside the city of San Antonio, Texas, is a natural cave. About 50,000 years ago it was the lair for a family of saber-toothed cats. The floor of the cave is generously strewn with many bones from their victims. Their favorite and most accessible food must have been young mastodons, because almost all of the bones were immature specimens of these massive creatures. Intermingled are a few bones of saber-toothed cats. Those who died were always included in the menu of those that remained.

No animal prints exist to suggest how the huge ice age cat managed to separate young prey from their enormous mastodon mothers. The answer may have been found by animal behaviorists researching African wildlife. Frequently they observed that a lone lion stalking a herd of elephants would keep a watchful eye on the very young. A baby elephant, inexperienced in self-preservation, often wanders playfully away from the watchful eyes of its feeding mother. If its wandering takes it near the concealed lion, its fate is quite predictable. The attacking cat rushes in, grabs the young elephant, turns it over on its back, and tears at the soft underside, wounding it fatally. Then the lion retreats into the bushes. By the time the furious mother elephant rushes to the aid of her infant, she is usually too late. She may linger for a while near the dead infant, but when she leaves, the lion carries off its prize.

Clearly the saber-toothed cat could have captured its young prey in the same way. Baby mastadons were doubtless abundant, easy prey and provided a generous meal for the pride.

This practice is quite old among carnivores, as evidenced by a finding near Camp Verde, Arizona, in 1958. A geologist was busily mapping the geologic rock formation known as the Verde Lake Beds when he uncovered an area that appeared to have been a large water hole in earlier times. Several hundred fossilized footprints, mainly of deer, horses, camels, and wolves, were impressed into the rock. The animals that came to this water hole to drink left their footprints in the area, which was then soft mud. The weather on that particular day must have been warm and dry, because all of the prints were baked by the sun. They hardened and were buried by the subsequent deposition of silt and sand. Eons later they were exposed near the surface by subsequent erosion.

The geologist was dazzled by the abundance of fossilized footprints. Of particular interest were four sets of adult mastodon footprints that emerged from the water hole. The prints measured nearly 12 inches in diameter, identifying them as prints of gigantic beasts. Directly in the middle of one set of adult tracks were the prints of a baby mastodon. These were doubtless the tracks of a mother mastodon followed by her infant.

However, a set of tracks of a saber-toothed cat was also present. These tracks approached the baby tracks at a slight angle and curved until they were superimposed directly on the tracks of the small mastodon. It takes very little imagination to re-create the scene. The baby mastodon, lagging behind the mother, was being stalked by the carnivore. The mother was no doubt oblivious of the danger to her offspring. The diagonal tracks indicate that the cat was closing in during its stalk. When it positioned itself directly on the tracks of the baby, the evidence is clear that it was preparing to attack. The stalk was over and the huge saber-toothed cat charged. This happened one day seven million years ago.

Ostrich Massacre

In birds, the gizzard is a highly muscular grinding and crushing organ that is part of the alimentary canal. It lies directly behind the organ that secretes gastric juices. Food is mixed with the gastric juices and passed into the gizzard, where it is ground up. The cells lining the gizzard produce a tough horny layer that aids in grinding. The bird will ingest small stones or pebbles (depending on its size) that also aid in grinding the food it swallows.

Some birds perform spectacular gastric feats with their rock-lined gizzards. Ducks and geese scarf up hard nuts, grains, and even live clams; chug them down; and crunch them up with the gizzard's lining of rocks. Clamshells, acorns, and corn kernels are all equally cracked into small pieces by this gastric mill. Pigeons also do very well. Their gizzards are especially tough and contain horn-covered "teeth" that grow from inside the lining. Even the hardest of tropical nuts are swallowed whole, passed into the gizzard, and cracked with an audible thunk.

Large birds, such as the ostrich in the wild, are selective in the stones they swallow. They seem to prefer rocks rich in quartz, and they wander the countryside in search of such potential "teeth." A large ostrich can carry around as much as a double handful of these strong, hard gastric tools.

For the ostriches that inhabited the plains of southwest Africa, choice of stone almost became their undoing. About a century ago a hunter shot a wild ostrich. In preparing it for the evening meal he cut open the gizzard and found several pure gem-quality diamonds among the stony contents. He set out early the next morning to hunt diamond-bearing ostriches. To keep such a find quiet is just about impossible, and word spread quickly. Within a week there was a grand rush onto the plains, and the slaughter began. Prospectors killed the defenseless birds by the thousands. Not all the victims contained diamonds, but some were fantastically rich; in one bird's gizzard 63 diamonds were found.

The ostriches were hunted almost into extinction in this part of Africa. The killing stopped only when too few survivors were left to bother with. The ostrich population has grown since the massacre, but an occasional potential diamond "mine" is still illegally brought down by the ever-present poachers.

Tropical Antarctica

It had been raining steadily for several days, at times rather hard, and flooding had become quite common. The thick soil on the side of the hill was now supersaturated and beyond its limit of stability. Suddenly, without any warning, the soil gave way and slid rapidly downhill. The thick, soupy, all-encompassing mudflow carried branches, small trees and shrubs, and rocks of many sizes, inundating everything in its path. With a roar it poured down the side of the hill and onto the adjacent floodplain, where it hit with enough force to knock over a stand of trees, burying them in the mud.

Such an event may have happened eons ago. The only certainty is that a small stand of trees was knocked over and killed by a mudflow. Radiometric dating showed that this happened about 260 million years ago. Fifteen mineralized stumps were preserved as fossils that scientists would find in 1992. Presumably the stand was a remnant of a much larger forest of trees. The saplings ranged from 3.5 to 7 inches in diameter; they were a well-known genus of seed fern called *Glossopteris*. Unlike true ferns, *Glossopteris* had seeds instead of spores; they were often treelike, and all are now extinct. Throughout the area the scientists found the tongue-shaped imprints of fallen *Glossopteris* leaves.

Deciduous trees such as *Glossopteris* are an indicator of a warm climate, as is the absence of frost rings. Growth rings of samples from the collected stumps contained none of the ice-swollen cells or gaps between cells that arise when the growth of a tree is disrupted by a seasonal frost. Scientists have concluded that these trees were rarely, if ever, exposed to temperatures below freezing. This is quite incredible, considering that they were discovered in rocks at an altitude of 7,000 feet on Mount Achernar, in the Trans-Antarctic Mountains. This mountainous area just a few hundred miles from the South Pole, was a floodplain millions of years ago when the plants lived there. Obviously, during the Permian Period 260 million years ago, deciduous trees adapted to a warm climate and grew in abundance in what we know today as Antarctica.

In the early days of geologic research, *Glossopteris* was an insufferable nuisance to earth scientists. Its fossilized remains were found unexplainably in the rocks of widely separated landmasses that included India, Africa, Antarctica, South America, and Australia. For several decades geologists envisioned sunken land bridges even though they found no geologic evidence to confirm pathways over the immense water barriers.

Glossopteris fossils are surprisingly homogeneous. Twenty species of leaves found in Antarctica are common in the rocks of similar geologic age in India, located north of the equator and half a world away. Seeds, much too large to be windborne, could not have blown across thousands of miles of open sea. Nor could they have floated across vast oceans. So how could a uniform land flora be dispersed throughout a hemisphere? The answer was finally provided by the application of plate tectonics, or continental drift. If the seeds of *Glossopteris* could not make the five-continent trip, the landmasses themselves had to move. Evidence has confirmed that, during the time these plants grew on five continents, the oceans that now separate them simpy did not exist. In fact these landmasses were joined in a single supercontinent known as Pangaea ("all earth").

The most startling, unlikely environment for *Glossopteris* is cold, windy, formidable Antarctica. During the Permian, however, Antarctica was positioned quite a distance from the south polar region, and its climate 260 million–240 million years ago was temperate to subtropical.

During the Jurassic Period, about 140 million years ago, Pangaea began to split apart. The landmass that became Antarctica drifted to its present geographic position, and India began the long voyage northward, to eventually collide with and become part of the Asian continent. The still-rising Himalayas and frequent earthquakes are evidence that this collision is still in progress.

Looking at a modern world map, one can easily see how South America would fit into the contours of western Africa. They resemble giant jigsaw puzzle pieces. Geologists have authenticated these connections by correlating the rock types and structures from Brazil into West Africa. They are identical and continuous and contain many other fossils of the same flora and fauna. To complete the puzzle, Antarctica, backed by Australia and with India to the north, fills in the space around the East African border.

Of any place on earth, Antarctica has shown the most dramatic evidence of climatic change. Today no true land animals live there, and only lichens, algae, and puny grasses grow where vertebrate animals used to roam through dense forests. But however strange a tropical Antarctica may seem, a greater challenge to the imagination would be a look at the Sahara Desert during the Ordovician Period about 500 million years ago. As landmasses shuffled for a more comfortable position on the planet, the land now known as the Sahara Desert surrounded the geographic South Pole!

But then, no matter what the temperature, barren wasteland is still barren wasteland.

Hairy Apes and Humans

In 1922, from his experiences at sea, Eugene O'Neill summoned up images of a stoker on an ocean liner who would become Yank in *The Hairy Ape*. With long, powerful arms and a bared, hairy chest, Yank is at home in the hold of a ship as leader of the stokers. In his search for meaning, Yank condemns capitalistic society and the people who seem to be in control but have no value. He and his coworkers, grimy people who eat coal dust and drink heat, are the ones who run the whole works; when the stokers start something, the whole world moves.

Yank's one-dimensional life has validity until the rich, pale, white-robed daughter of the chairman of the board visits the bowels of the ship to see how the other half lives. When Yank suddenly confronts her in the light of open furnace doors, she screams, "Filthy beast, hairy ape!" then shrinks away and collapses whimpering. Yank is thoroughly sabotaged when the attributes of which he was most proud are attacked. He tries, without success, to find his place in the world and finally discovers kinship with a gorilla that kills him in an embrace.

Most people, just as Yank, would not feel complimented by being described as a hairy ape. After all, we pride ourselves on having progressed from our simian ancestry. Humans, from all appearances, have lost most of their body hair and continue to move farther from the apes. The surprise is that the human body has the same number of hairs as are found on the bodies of apes; it is just finer, shorter, and lighter. Like humans, apes become gray or silver with age and suffer hair loss, particularly on their heads.

Great apes, not to be outdone by humans, have their own individual sets of fingerprints. Moreover, not to be outdone by dogs, no two have the same noseprints. Zoologists prefer to identify gorillas by their noseprints, but then they have to coax the gorilla to submit to having its nose smeared with ink and pressed onto paper. We lose more zoologists that way!

Conquest Through Religion

The Aztec, despite their humble beginnings, produced one of the greatest civilizations of Mexico. Originally they were nomadic warriors from northern Mexico who settled about A.D. 1200 on a small, heart-shaped island in Lake Texcoco and created the city of Tenochtitlán. From here, as an amphibious community dependent on the lake, they adapted sandbanks, marshes, and a network of islets for the purposes of living. These warlike people conquered central Mexico and imposed worship of their gods on the nearest villages on the mainland. Doubtless the conquests were quite barbaric, but ruthless savagery seemed appropriate to their religious beliefs. The Aztec reached their greatest power and development under Montezuma II, who ruled from 1502 to 1520. With the coming of the Spanish conquerors, his downfall, and that of his people, was abrupt and final.

Perhaps the collapse of the Aztec civilization resulted from the beliefs basic to their religion. According to legend, in the midst of darkness one of the gods hurled himself into the flames of a great fire. He

emerged transformed into the brilliant sun. But the sun remained immobile and would not move without "fuel," which of course was blood. The other gods sacrificed themselves so that the sun, deriving life from their death, might begin moving. This symbolized the Aztec belief that life can come only from death; there is merely a transformation of energy.

Because the sun died every night, it had to be resurrected with human blood. Only this "precious water" would give it the strength to be reborn and to move across the sky. Human sacrifice became imperative as the sole means whereby humanity could survive at all. Without the steady flow of blood, the sun would stop and the entire world would plunge into darkness and death. About 15,000 humans a year were sacrificed to the sun god Huitzilopochtli. However honorable sacrificing oneself may have been, most of the victims were prisoners taken in the frequent wars, which were often started solely to round up victims for sacrificial rituals.

The Aztec war machine, always seeming to be in the act of conquering, was really geared more to the capture than to the slaughter of the enemy. In the military hierarchy, rank and honor depended upon the number of captives taken; a man who took no prisoners carried no esteem, even if he were a prince. This obsession partially accounts for their dismal performance in the open field against the relatively small band of Spaniards, who sought to kill with their swords. The Aztec weapons were more suited to wound; the rationale was that a dead enemy could not be sacrificed to the gods.

Aztec wars of conquest and prisoner-taking were an almost constant fact of life, and the gods were kept satisfied with the flowing blood of sacrifice. The greatest recorded sacrifice of all time took place in the year 1487, in the reign of Ahuitzotl, to inaugurate the Great Temple, which of course had to have its start in blood. The emperor, together with two allied rulers and his leading official, led the way, accompanied by throngs of priests. Everyone, priests as well as victims, was dressed as a god. The captives who were to be sacrificed, duly painted and feathered, formed interminable lines that stretched along the main causeways into the city. The performance had an enormous captive audience since the people of neighboring towns were ordered, on pain of death, to come to the capital to watch the proceedings. No one was exempt from attending the ritual, and not a single man, woman, or child was seen in the streets of their hometowns.

Records confirm that the sacrificial slaughter lasted four complete days. When Ahuitzotl and his royal colleagues wearied of gashing open the victims' breasts and tearing out their hearts, priests and other lesser dignitaries took up the knives in their place. So covered was the temple with fresh blood that it appeared to have been painted red. According to tradition, the victims numbered 80,400—a figure repeated by several sources.

Many daily activities of the Aztecs were sacred in nature, and human sacrifice in their daily life was common. One of the more violent games, played only by dignitaries and warriors, was tlachtli, played with a hard rubber ball on an I-shaped court. The object of the game was to knock the ball into the opponent's end of the court, much the same as in modern volleyball, except without using hands or feet to move the ball. Teams could also win the game by making the ball pass through either of two stone rings set on the sidewall of the court. The rings were 20 feet above the ground and just large enough for the ball to pass through. On the rare occasion when a goal was achieved through the rings, the person who scored, along with his friends, was allowed to confiscate clothes and possessions of any spectators they could catch. Most often they would attempt to catch a rich noble, the only person able to repay them for what they had endured.

Since the ballgame was fairly dangerous, players were covered with a protective, armorlike leather garment. Still accidents were quite frequent, with crushed ribs or chest and broken limbs common. Occasionally a player unable to rise from the field at the end of a game would simply bleed to death. And a defeated player just might have his more successful rival assassinated. In either case, one of the bloodthirsty gods would profit from the sacrifice.

One of the more unusual customs practiced by the Aztec was their selection each year of a very handsome young man to be sacrificed to their chief god Tezcatlipoca. For 12 months the man was treated royally as if he were a king and allowed every possible luxury. He was taught to play the flute, an important skill for royalty. He ate the same foods as the king and was wined, fed, and doted upon constantly. His every whim was taken seriously and satisfied immediately. He spent the last month with four beautiful girls of his own choosing. At the end of the month he led a procession to the temple of Tezcatlipoca. As he lay on the sacrificial altar, four men held him down while a priest, using an obsidian knife, cut open his chest and tore out his heart. His death would ensure the continued life of the god.

Historians are sure that the all-powerful religion of the Aztec was the unifying ingredient in their lives. The blurred relationship between life and death made human sacrifice infinitely acceptable. Ancient predictions of the future were common knowledge to the Aztec. For example, the most powerful wind god, Quetzalcóatl, was predicted to return from exile in 1519. He was fair-skinned and bearded, and would be riding on an animal (which turned out to be a horse) that only the gods could ride.

By an extraordinary coincidence Cortez unknowingly fulfilled the ancient Aztec prediction. The bearded Spaniard, bent only on conquest and plunder in the name of his country, arrived on the Atlantic coast to begin his invasion of Mexico in 1519, just as predicted. When he appeared, all the chiefs and priests fell on their knees in awe and hailed him as the god Quetzalcóatl. Even the great Montezuma thought Cortez was the returned wind god. No real resistance would be permissible against a powerful god.

The Aztec were so thoroughly beaten that their descendants were never able to rise from the disaster, in which all authority, beliefs, and religion collapsed. All that remained of this brilliant civilization was the peasant. With his hut, his field of maize, some poultry, and limited hopes for the future, he alone survived. Historians, both native Aztec and Spanish conquerors, have ignored this peasant, who, now as then, continues to lead his patient, laborious life in nonthreatening obscurity.

Galveston's Day of Infamy

"Red sky at night sailors delight;
red in the morning, sailor take warning."

Old mariners' adage

At the beginning of the 20th century, Galveston, Texas, was already a major deepwater port. Steamers, freighters, and sailing ships regularly docked at its wharves. The city was bustling, wealthy, and proud, boasting a population of over 25,000 people. The beaches were lined with substantial homes of merchants, and along lush, tree-lined boulevards stood the ornate mansions of the rich. There were numerous large churches, an orphanage, a convent, hospitals, an army fort, and many public buildings. Galveston was definitely a city on the move, and many believed it was destined to be the most important city in the South.

But Galveston had a problem. Its ideal location as a deepwater port was on a barrier island that partitioned Galveston Bay from the Gulf of Mexico. The island is only three miles wide with an elevation of less than 20 feet at its highest point. Galveston lay ominously exposed. The city had been subjected to hurricanes during the previous century, but rebuilding had kept it up-to-date.

On the pleasant afternoon of September 7, 1900, Galveston was still playing host to many people who sought to escape the summer heat of the nation's interior by vacationing along the Gulf Coast. No one in Galveston imagined that within 24 hours the city would be almost completely destroyed, a victim of one of America's worst storm disasters. Sailors were reporting that their ships had come through "sheer hell." Somewhere out there was a storm of gigantic proportions. Weather officials in Galveston knew on September 1 of a hurricane in the eastern Caribbean. But the first official telegraph from Washington, D.C., was not received until 4 P.M. on September 4. It said simply that a tropical disturbance was moving northward across Cuba. In those days, before the time of aircraft patrols, radar, and satellites, following a storm was impossible.

On September 7, chief forecaster Isaac Cline and others at the weather bureau noted the falling barometer, a change in wind speed, the rapidly rising humidity, and the change in sea waves beating on the Galveston beaches. Then, acting on information circulated by the weather bureau, the local newspaper published a forecast for the next day: hurricane. Almost everyone ignored the forecast. In fact, many people took excursion trains to the city.

On September 8, a Saturday, a large number of sight-seers arrived, attracted by the raging surf. Cline climbed to the roof of the weather office and hoisted storm-warning flags. Convinced that his city was lying in the path of the first hurricane of the new century, he headed for the beach area to warn the people to move immediately to higher ground. Again his warnings were ignored.

As Saturday afternoon wore on, conditions grew worse. Hurricane flags were snapping in the wind and ripping from their masts. Yet people thrilled to the sight of enormous waves that arced up out of the sea and fell on the beach with a thunderous roar; a few daring souls splashed about in the surf.

A regular beach attraction was the Pagoda, a structure shaped like an oriental temple that covered about two blocks. The crowd watched amazed when the waves effortlessly demolished it in front of their eyes. The great building sank into the water and in minutes became a disordered mass of floating timbers. Finally people began to recognize the severity of the storm and raced for shelter.

The waves first surged onto the houses at the beach, splintering them and shoving debris inland. Then they gushed into the street, lifting and breaking house after house. Soon the entire city was racing away from the shore. Keeping close together, crowds of people herded toward the middle of the city as the wind speed rose to 120 miles per hour.

Those who ran before the wind faced a new terror. Darkness was all around now, and they could not see the deadly slabs of slate that flew through the screaming air. After the great Galveston fire of 1885, city fathers had recognized the value of slate roofs to halt the spread of fire and make their city safe. They had rewritten the building code to make wooden roofs illegal. Now these stone slabs, torn off roofs, were as deadly as meat cleavers.

When the water first crept into houses, people retreated to the upper floors. As the waters continued to rise the people ran out of floors to ascend to. After he retreated into his home with his family, one survivor related, the house started to shudder as a battering ram of huge timbers from wrecked houses crashed against the walls. The building began to come apart, slowly at first. Then with great suddenness it disintegrated and collapsed into the churning waters.

Gigantic waves ripped away wharves and set ships loose to drift helplessly before the onslaught of wind and waves. Occasional flashes of lightning revealed the eerie picture of huge oceangoing steamers wallowing in a mass of floating houses. Other ships hit rocks and sank immediately, with a heavy loss of life among the crews. Some ships were pushed so far inland that when the storm ended, they were sitting in the middle of Galveston or in a farmer's field.

In the face of 120-mile winds, chimneys flew apart, church steeples collapsed, brick walls disintegrated. Refugees clinging halfdrowned to the wreckages of their houses were mauled by the storm or buried in cascades of bricks and wooden debris. Waves washed people off roofs, while tossing beams knocked children out of the grip of parents. Swimmers gave up and sank beneath the waves. The air was filled with flying debris, the howling noises of the winds, and the terrified screams of the dying.

One man rode out the storm by climbing inside a handy barrel. He was able to secure the lid so that the barrel was watertight. All night, as he bounced on the waves, he could hear the wind, feel objects hitting his barrel, hear the screams of drowning people. Another family survived by tying themselves in a tree. Trapped by water and waves they climbed a tall tree near their house and lashed themselves to the trunk. All night hurricane winds swayed the tree, cracking off limbs and ripping off most of their clothes. When the water subsided wind-driven sand flayed their skin. The entire family became ill from swallowing so much saltwater. They were among the fortunate few who survived being tied to a tree. Most others were drowned as the water rose over them or were cut to ribbons by flying objects.

Bolivar Lighthouse offered refuge from the storm. A virtual mob crowded into the lighthouse to take shelter in the tall structure with heavy stone walls. People crammed in on top of other people and, on the steps of the spiral staircase, sat several deep on the laps of those already seated. After the last person got in, the heavy door at the foot of the stairs was closed. Waters quickly rose around the base of the lighthouse and the door was soon deep under water. It was pitch-black inside. Sitting in cramped positions for hours, many suffered from muscular pains, and the air became foul. People sobbed and became maddened by their pain and tormented by their thirst. Many wished they were dead.

All survived the ordeal. When the waters receded, they opened the door and ran into the sunlight. The first sight to greet them was a mass of dead bodies that lay all around the base of the lighthouse.

The Ursuline Convent, a large stone and brick structure, covered four city blocks. Around the convent was a 10-foot-high brick wall that protected the convent buildings throughout much of the storm. When the wall finally crumbled, water rushed through as it would through a broken dam, carrying in cargoes of trees, crates, rooftops, corpses—and living humans. The nuns immediately went to work to rescue them. Leaning from windows and using poles, they dragged people inside. One woman about to give birth when the storm started had managed to get into a large trunk that carried her bobbing up and down over the raging waves. She eventually floated to the convent where the nuns pulled her in through a window. She immediately gave birth to one of eight infants delivered in the convent that night.

Another man rescued by the nuns was stark naked. One nun gave him a habit to wear; dressed as a nun, he immediately went to work. A few dazed people were puzzled by the bearded nun helping them. One hurricane victim was even more astonished—it was his wife, who had thought he was dead. He, in turn, had thought she was dead.

Protected during the worst of the hurricane by the surrounding wall, the Ursuline Convent, though heavily damaged, survived, as did those lucky enough to have been rescued by the nuns. Not too far away, the Old Women's Home caved in.

As rapidly as the great Galveston hurricane had built to a climax, it retreated. After midnight the winds began to die down, and by dawn they had lessened considerably. When the sun rose there was only a stiff breeze—and a scene of utter devastation. Half of the buildings in Galveston were completely destroyed; almost none had escaped heavy damage. Amid ruined, smashed houses lay the pilings of broken bridges, smashed boxcars, ships, telephone poles and wires, dead animals, and thousands of corpses. No contact could be made with the world beyond Galveston. There was no immediate way to build a fire or cook a meal, and no water to drink.

Within a few days the rest of the country began to respond, and into the stricken city came all manner of help, including martial law. A most urgent problem was disposal of the dead, which were literally everywhere. The stench of thousands of decaying bodies eventually made breathing almost impossible. To avoid such epidemics as often follow a disaster, officials decided to dump the corpses at sea. Although thousands of people wanted to identify their dead relatives and friends, the need to dispose of them was immediate. So the bodies were loaded onto channel barges, towed to sea, and dumped. Unfortunately corpses float. The wind blew them right back to Galveston. From piles of broken boards, workers built pyres on which to burn the corpses that washed back on land. For the survivors this was the ultimate agony.

By official count as many as 8,000 people died in the terrible hurricane. This is by far the largest death toll in a U.S. national disaster.

Nothing comparable has happened in the United States before or since. Not even in Galveston. Within a few years a seawall was built, so massive and strong that no hurricane of the future could devastate the city. Nor did the city want to skulk behind a high, prisonlike barricade. The entire city was raised to the level of the 17- to 20-foot wall. Since then the residents have been able to look down on a less-threatening sea. Nevertheless, when they hear warning of a major hurricane, the citizens turn toward the continent across Galveston Bay and head inland.

BOINGGG!

All owners of furry or feathered pets will at some time live with fleas on the premises. These tiny insects feed on the blood of birds and mammals, including any humans that might be around. Unless the animal is badly infested, fleas aren't easy to see. They avoid the sparse-hair regions and head for home in the forest of thick hair.

Occasionally a flea will fall or jump off a pet and be stranded on the carpet. Since flea food—fresh blood supplied by the host animal—is not plentiful on a carpet, a human that walks in the room is a welcome source of food for the wandering flea. The new host often isn't aware of having been selected as a flea meal and will go to bed that night not knowing of a new tenant residing somewhere around the ankles. In the morning numerous itching welts give evidence of the presence of a flea. By this time the flea is usually back home in the forest of hair of its favorite pet, much pleased to be enjoying its usual host.

Fleas, natural-born jumpers, confine their leaps to hopping from place to place on the host. They are built conveniently narrow so that they can walk through coarse hair. Jumping ability is a safety valve in case they need to move from here to there rapidly. Nevertheless, for an insect less than one-eighth inch (3 mm) long, the flea's ability to make a long jump of 13 inches (33 cm) and a high jump of 8 inches (20 cm) is quite spectacular.

The jumping power of the flea comes both from strong leg muscles and from pads of a rubberlike protein called resilin located above the flea's hind legs. To jump, the flea will crouch, thereby squeezing the resilin, and then relax certain muscles. Stored-up energy from the resilin works like a gigantic spring that launches the flea. Perhaps *launching* is too mild a word to describe the action of a species that can jump 150 times its own length horizontally or 80 times vertically.

To reach the same distance as the flea, proportionately, a human being would have to spring horizontally the length of two and a half football fields or vertically the height of a

60-story building—in a single bound, from a stationary sprint position. The best Olympic long jumper, at 29 feet (about 10 yards), would have to add 240 yards to the previous score. An Olympic jumper might appeal for help by pleading, "Oh, to be a lowly flea!" Otherwise the Olympic gold might go to the lowly flea.

Neanderthal Bedouins

Neanderthals are usually associated with the cold glacial environments that existed during the last ice age. But this was not always true. They were quite widespread, and some groups even existed in hot, parched lands.

Primitive humans usually stayed within walking distance of water, but a certain group of Neanderthals managed to overcome this restriction by utilizing water vessels. These vessels were not pottery bowls, but ostrich egg shells, which made perfect canteens. The modern African bushman often employs an ostrich egg shell as a water vessel during migration or on the hunt.

The evidence of the Neanderthal desert versatility was recently exposed in the dry, sun-baked Negev region of Israel. Here excavation of a Neanderthal site revealed a number of stone tools along with several intact fossil ostrich eggs. Each egg had a small punctured hole that doubtless served as a mouthpiece.

The Time Machine

Is travel through time really restricted to the realm of science fiction? No, not entirely! Centuries ago man invented an instrument that would permit Earthlings to witness events as they happened and things as they appeared in the very distant past. This window into days and eons gone by is a well-known, much-used instrument called the telescope.

Contrary to popular opinion, Galileo did not invent the telescope. A Dutch optician, Hans Lippershey (1570–1619), is usually credited with its actual invention, although other people, long forgotten, had experimented with lenses and almost certainly had discovered the principle before Lippershey assembled lenses into a telescope. However, Galileo very likely was the first to view the magnificence of our solar system. And when he turned his telescope toward outer space, he was among the first time travelers.

Galileo got himself into a lot of trouble because of his telescopic revelations. Back in the late 16th century, the only acceptable belief was that Earth was the center of the universe. It did not move because every celestial body, including the sun, revolved around Earth. This viewpoint was first challenged by Nicolaus Copernicus in 1530 but for obvious reasons was not published until 1533, while he lay on his deathbed. Copernicus, ignoring the wisdom of Ptolemy and the ancients, had displaced man from the center of the universe.

To agree with Copernicus was unhealthful in those days. One Giordano Bruno could attest to that; for he was burned at the stake in Rome in 1600 because of his heretical views that Earth, "this center," moved. Shortly afterward Galileo was seeing unseemly sights through his telescope: craters on the moon, satellites of Jupiter, the ring of Saturn. Much of what he saw proved that the earth was not the center of the universe. After writing his *Dialogue on the Great World Systems* in 1632, he too found himself up before the inquisitors. Rather than sacrifice his work and himself, he recanted and confessed his error in confirming Copernicus. According to legend, after recanting he quietly muttered, "yet it moves!" His last 10 years were spent in internal exile, under close watch, just as a modern parolee reports periodically to the parole officer. Galileo's name was cleared a few centuries after his death. In 1979 the Pope declared that the Church's verdict on Galileo should be reexamined.

Light travels over 186,000 miles per second (six trillion miles per year). The light that reaches us from the sun left its source eight minutes earlier to make the 93-million-mile trip to the earth. On a clear dark night, an observer can see, even with the naked eye, a sky full of twinkling stars. In most cases the light one sees left the star many eons ago. Even within our solar system, the light reflected from Pluto takes five hours to reach Earth. At 24 trillion miles our nearest star, Alpha Centauri, is a little farther than four light-years away. The astronomer's telescope is, in effect, a modern time machine, for it reveals remote areas of the universe as they appeared up to billions of years ago. If there are planetary systems around Antares in Scorpio and Betelgeuse in Orion, their inhabitants can see Columbus discovering America in about 15 years.

In 1975 a very distant galaxy was discovered, over eight billion light-years from Earth and still receding. The light seen by the astronomer left this galaxy eight billion years ago to make the 48-nonillion (30 zeros) mile trip to the earth. Since Earth is only about 4.6 billion years old, the astronomer observing this galaxy sees it as it appeared nearly 3.5 billion years before Earth was born. Unfortunately the viewer from the distant galaxy training a telescope toward our solar system will have to wait 3.5 billion years to see an event of great importance on our planet—its birth.

If you ever give a telescope to a travel or history enthusiast, a perfectly appropriate label for the package would be "Enclosed: one time machine."

CHAPTER TEN

"The Tempest"

Hurricanes as we know them today were forming and smashing against the islands of the Caribbean and the eastern seaboard of North America long before Columbus made his first visit to the Americas. This type of storm was called a tempest until the Spanish and English explorers combined the names for thunder, lightning, and storm gods of native tribes throughout the Americas into the word *hurricane*.

Many are the stories of trouble and near-mutinies that Christopher Columbus faced with his crews in 1492. Had the navigator from Genoa known he was sailing straight into a region of violent storms near the season of their peak annual performances, he might have been willing to turn back. Possibly 1492 was a freak year in which no hurricanes blew, or perhaps the divine hand of Providence steered storms around the little fleet. Whatever the reason, not once during the months that Columbus spent on that first voyage to the New World did he encounter a hurricane.

The only time he experienced hurricane-related conditions was during his approach to the eastern Bahamas. Here the fleet ran into high seas with no wind; the *Niña*, *Pinta*, and *Santa Maria* were almost surely sailing through storm tides piled up by a hurricane. It had merely recurved sharply behind them and wasted away in the North Atlantic.

Columbus appears again to have been spared a true hurricane in 1494. Then, on his final voyage, Columbus plodded along the coast of Central America, hoping that the turn of each new headland would provide the coveted passage to the Far East. Full-blown hurricanes are rare in the lower Caribbean, but this time Columbus managed to meet one head on. The torment of the storm was recorded in his log, which contained these observations: "The tempest arose and worried me so that I knew not where to turn; ... eyes never beheld the seas so high, angry, and covered with foam. ... Never did the sky look more terrible. The people were so worn out that they longed for death to end their terrible suffering."

Records of hurricanes that occurred in the 17th and 18th centuries were surprisingly accurate and complete. While the New World was being colonized, knowing the habits of hurricanes, when and where they were likely to strike, was important. One particular 17th-century hurricane is important because it appears to have provided William Shakespeare with background material for *The Tempest*, his last contribution to the stage.

On June 2, 1609, a fleet of nine ships set sail from Plymouth, England, bound for the colony at Jamestown, Virginia. Aboard one of the ships, the *Sea Adventure*, was the new governor of Virginia, Sir Thomas Gates. The fleet plodded wearily across the Atlantic for almost two months without incident. In late July a hurricane passed through the Bahama Islands and recurved out into the ocean. On July 25 it struck the fleet; almost immediately it sent one of the ships to the bottom and totally scattered the others. One by one, seven of the nine ships straggled into the harbor at Jamestown. When several weeks had passed and the *Sea Adventure* failed to arrive or be heard from, it was believed lost, along with the new governor and at least 150 others.

The *Sea Adventure* had not sunk but had become completely separated from the fleet. In its attempt to make a landfall on the island of Bermuda, luck ran out; the ship went to pieces on a reef. Most of the crew and passengers managed to get ashore safely and, for the next nine months, lived as castaways on the island, with the wreckage from the ship and the natural bounty of the islands to sustain them. They built two small vessels and finally sailed safely to Jamestown.

Sir Thomas and some of the survivors returned shortly to England, and the story of their adventure became the talk of London. Two returning survivors, Silvester Jourdain and William Strachey, each published a pamphlet documenting the events of the voyage. Strachey's

True Reportory of the Wracke describes vividly the dreadful storm, rebellious castaways, compassionate governor, treacherous natives—the stuff of a good drama.

Scholars have little doubt that the 1610 manuscripts became William Shakespeare's inspiration for *The Tempest*. Many parallels of the shipwreck and the Bermuda experience appear in the play. Although the uninhabited island is not identified as the New World, its climate and topography coincide. Other noteworthy parallels are the type of storm and analogous characters. Sir Thomas Gates is, of course, Prospero the Duke of Milan, and the Native American is Caliban, the deformed slave, who alludes to making dams for fish and finding cedar berries in the water. The usurpers and drunken sailors are colonizers who plan to exterminate or enslave the natives.

Shakespeare took many liberties in retelling the story, changing all names and places to protect the plot. The result has been the most memorable hurricane in history or legend.

Sit-Down Strike

From the building of the step pyramid 4,600 years ago through the Old and Middle Kingdoms, Egyptian pharaohs were buried in pyramids. About 90 of these monuments rose from the sands of Giza and other cities near Memphis. Despite the ingenuity of the builders, tombs were almost always plundered. As a result Pharaoh Thutmose I decided, some 3,500 years ago, to create an underground tomb for himself in what came to be known as the Valley of the Kings. It was located on the west bank of the Nile, about 400 miles south of the pyramids, accessible to the capital city of Thebes and yet concealed from possible robbers.

For the next 500 years, through the period of the New Kingdom (1567–1085 B.C.) all the pharaohs were buried in that same Valley of the Kings. As far as is known each of the tombs was eventually robbed, but less systematically and less quickly after the monarch's burial.

Thousands of fragments of limestone and pieces of broken pottery covered with drawings and hieratic writing (hieroglyphic shorthand) give detailed information concerning conditions under which tomb excavators and builders lived and worked. Contrary to popular belief, work crews for the royal tombs were not slave laborers who eked out a miserable existence under brutal working conditions; they were a permanent company of skilled artisans. The work of burrowing into the rock and constructing tortuous passages, hidden doors, and false chambers was carried out by specialized craftsmen. Artists and draftsmen decorated the interiors of the tombs.

To ensure secrecy of the tomb's whereabouts, the tomb builders with their families were housed on a barren site in the desert in a specially constructed village, Deir el-Medina. Each single-story house consisted of four rooms, and all were lined up facing directly onto the street. Although none had its own water supply, a public tank stood outside the main gate of the enclosure wall. Servants were provided by the government to grind grain, do laundry, and mend pottery.

Records recovered and deciphered provide incredibly detailed information. Each workman's name and duties were listed, and a timecard noted his arrival at work. The progress of work on each tomb was care-

fully charted. These records were quite specific: on one fragment it was noted that a workman was absent because he had had an argument with his wife. Other inventive excuses ranged from "eye trouble" and "brewing beer" to "embalming mother."

The men were organized into two working crews, each under a foreman, his deputy, and a scribe who kept careful records of everything that took place. They worked on eight-day stretches, during which they slept in simple huts close to the tomb under construction. Every 9th and 10th day was a holiday, at which time they went home to their wives and children in the village. There was also time off for festival days of the principal gods.

Workers were generally conscientious, and wages were paid in wheat for bread and barley for beer, issued monthly from the royal treasury. The men were also supplied with rations of fish, vegetables, wood for fuel, and body oil. The latter was in great demand by the men working in hot, dusty conditions. Occasionally the pharaoh himself would reward his skilled tomb workers with luxuries, such as meat, wine, salt, and Asiatic beer.

Isolated in the desert and unable to grow their own food, the villagers relied on the prompt delivery of supplies. This normally happened on the 28th day of each month, but occasionally the heavily laden donkeys failed to arrive on time.

In the 29th year of the reign of Ramses III, the supplies were several weeks late and nowhere in sight. Finally the workmen threw down their tools in disgust and made their way to the great mortuary temple of Ramses II. There, in an orderly fashion, the employees sat down and refused to return to work until the pharaoh had been informed of their desperate plight. A temple scribe, after hearing the men's complaint, ordered that they be given a month's supply of grain from the supplies allotted to the official scribes. The men staged further strikes over the next few months, until the backlog of monthly payments had all been delivered.

No record shows anyone being punished for daring to dictate terms to the pharaoh in this way. The evidence indicates that the rulers of ancient Egypt were less tyrannical and the workers less docile than history has sometimes led us to believe. (Hollywood has certainly helped in shaping this negative image of the great pharaohs.) The workers and the pharaoh knew their work was absolutely vital to the king. His journey to the next world could not be made unless his "house of eternity" was decorated, furnished, and completed in time for the gods to receive his earthy remains. It therefore seems reasonable that the king would treat the men well, even to the extent of meeting their collective bargaining demands after the first known "sit-down strike" in history.

Animal Forecasters

In 383 B.C., about two days before an earthquake destroyed the city, Helice, Greece, witnessed a mass evacuation of the city's rats, snakes, weasels, millipedes, and even worms. This animal movement was spectacular enough to be recorded by the Greek historian Diodorus, who quoted witnesses to the incident.

The Lisbon, Portugal, earthquake of 1755, which rang church bells in Sweden, is mentioned in the writings of German philosopher Immanuel Kant: "Eight days before the tremor the ground near Cadiz was covered with a multitude of worms that had crawled out of the ground."

On July 26, 1908, the night before a quake devastated Naples, Italy, great swarms of locusts crawled through the streets and into the sea. Oxen bellowed, sheep and goats bleated, dogs howled, and geese squawked noisily.

A rancher on Kodiak Island noted on March 27, 1964, that his cattle would not stay in their low-lying pasture. He didn't argue with whatever instinct took them to higher grazing land. When the earthquake came at 5:36 P.M. a great tsunami rolled over the lowlands. In that same great quake the Kodiak bears came out of hibernation two weeks early. Tracks showed that they didn't stop to feed as they normally would when awakening from their long winter sleep. Instead their tracks indicate that they left their rock caves and headed out on the run.

Few residents of Parkfield, California, will ever forget June 25, 1966, when the town was invaded by rattlesnakes. Why did the reptiles flee the dry, grassy hills nearby? The answer came two days later when the surrounding hills were shaken by a medium-to-heavy earthquake.

The night before the Sylmar earthquake of February 9, 1971, several police patrols reported independently to their dispatcher that enormous numbers of rats were scurrying through the streets of San Fernando, California. The police also received a number of complaints about incessant barking and howling of dogs for several hours before the 6:01 A.M. quake.

Although unusual animal behavior preceding earthquakes has been observed and reported throughout the world for hundreds of years, researchers have been reluctant to assign credibility to the phenome-

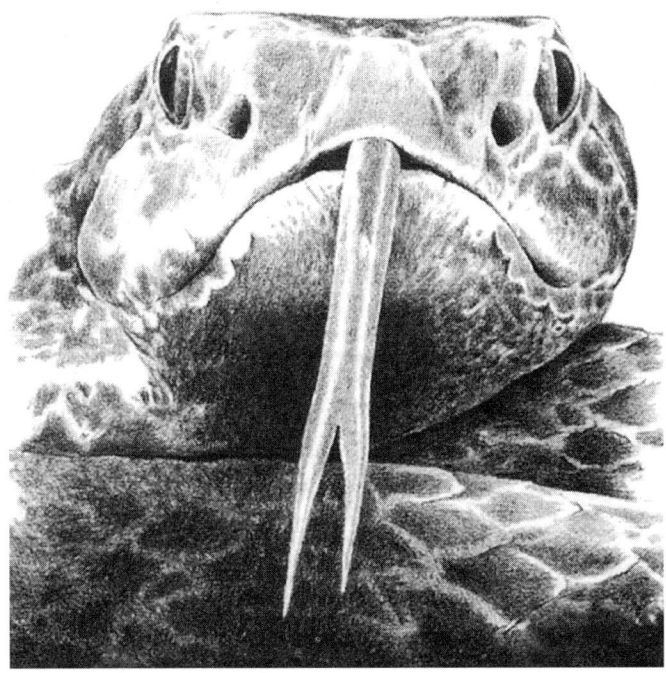

non. Reports of animal anxiety, often mingled with allegory and legend, were considered flights of the imagination, superstition, fantasy, and old wives' tales. But now the collected accounts of many centuries are finally being accepted as reliable information.

In June 1974, China's National Earthquake Bureau issued a warning that a critical earthquake could be expected near the city of Haicheng, in Lianoning Province near Manchuria, within one to two years. This prediction was the result of a four-year study of a 2,200-year history of earthquakes in the area. Their conclusion that geologic changes were imminent was supported by precise seismologic, geodetic, and geomagnetic measurements.

The citizens were mobilized into a large-scale prediction network. With the guidance of experts, volunteers were placed in thousands of industrial plants, schools, animal breeding institutions, and agricultural communes; others were trained to observe animal behavior. They were alerted to recognize other signs of an impending earthquake, such as changes in the water table, bubbling in water wells, sulfur smells in wells, increased radon gas, earthquake swarms, weird lights, or strange rumbles coming from the ground. The collected information was assembled in a central clearinghouse.

Many observers acknowledged that the most widespread, predictable, and meaningful signs of an approaching earthquake were the strange behaviors of animals. They would signal their fears loudly, leave their burrows, and try to break out of any enclosure. Observers noted also that several small earthquakes definitely signaled a forthcoming earthquake. By February 1975, these signs had become very real.

More than 20 species of animals showed unmistakable signs of fear and confusion. Large animals such as horses and cows became unmanageable and panicked, kicking, biting, and running away. Snakes came out of hibernation, crawled from their burrows, and froze to death on the snow-covered ground. Rats appeared in the open in large groups, so confused that they could be caught by hand. Geese flew into walls and trees, pigs bit each other and dug out of their sties, chickens refused to enter their coops to roost. Even trained police dogs would not obey their handlers, but howled and kept their noses to the ground.

Several small earthquake swarms were detected in the first days of February. This evidence, along with groundwater changes such as wells turning cloudy and gas bubbles in ponds, convinced the scientists that the anomalies in animal behavior were relevant. On February 4, 1975, evacuation of the Haicheng area was ordered and was in progress by 2 P.M. The earthquake, 7.3 on the Richter scale, hit at 7:36 P.M. Over 50 percent of the houses were destroyed. In this city of about a half-million people, tens of thousands would have died but for this successful prediction and evacuation. The few casualties were stubborn individuals who insisted that this was all poppycock; their houses collapsed on top of them.

Eight months of observations by the Chinese produced the following telltale signs of imminent seismic activity: (1) a dramatic rise in incidents of abnormal animal behavior, (2) distinct changes in the groundwater level, and (3) swarms of small earthquakes. Buttressed by each other, these observable events are precursors to a significant main event.

Much progress is being made in the long-range study of earth movements. Technology for constructing earthquake-resistant buildings, bridges, and physical facilities is coming of age. But short-term predictions provide the most effective measure of earthquake preparedness; they signal when to evacuate. For this, the early warning signs provided by animals are a most important feature in earthquake prediction.

A Most Dangerous Creature

If you were asked to envision the world's most dangerous creature, what would come to mind? Doubtless something huge and scaly with claws glistening, fangs bared, and hoofs flailing; with horns, stingers, world-

class poisons, or other built-in weaponry; and sporting an impenetrable armored coat.

Most of these features are absolutely unnecessary, for an animal equipped with neither weapons nor defenses is, if not the champion, at least among the front-runners of creatures that plague humankind—*Musca domestica*, the lowly housefly.

The fly, worldwide except in the Antarctic, is of all animals consistently the most dangerous to man. The diseases it transmits bring more misery and death to humans than could any imaginary monster. Having selected humans as its permanent companion, the fly carries over 30 different serious diseases to people all over the world: typhoid, tapeworm, hookworm, whipworm, yaws, diarrhea, amoebic dysentery, tuberculosis, ophthalmia, trachoma, septicemia, gangrene, conjunctivitis, or any disease caused by the millions of bacteria the housefly can carry around on its body and legs.

The fly's body structure and lifestyle guarantee a ready supply of disease-producing bacteria to redistribute among its human companions. Almost constantly on the move, the fly travels many miles in its 30-day adult life. During its short, active life span the fly, attracted to aromatic garbage, decaying bodies, and manure, constantly picks up microorganisms on its hairy body and sticky footpads. Then it hurries off to the equally fragrant food preparation and serving quarters of humans. The average fly carries about 1,941,000 bacteria on its body in clean communities, 3,686,000 in areas without sanitation. Most insect larvae compensate for the destructive activities of the adults, but not the housefly. Rather than hastening the disintegration of the waste matter on which they live, the larvae devour the organisms that break down refuse and retard the process of decay.

A "fly fact" even more horrific is their incredible rate of reproduction. A female housefly lays several hundred eggs, which hatch in 8–12 hours into maggots that become pupae and then a new crop of adult flies within a week or two. If the eggs of just one fly and all those of her offspring were allowed to hatch and grow with no casualties, the result would be, in about six months, over 5.5 trillion flies! They collectively would weigh over 80,000 tons and would carpet the earth to a depth of three and one-quarter feet. The meek would truly inherit the earth. Mercifully for the earth's inhabitants, only an infinitesimal number of fly offspring make it to adulthood.

As a result of the fly's prolific reproduction and short life span, its evolution is speeded up enormously. Within a few generations the fly develops immunity to almost every type of insecticide. This immunity is complete within a year or two. Attempts to control or wipe out the housefly with chemicals would poison much of the surrounding environment (as did DDT) before they could even make a dent in the fly population. Therefore, control of houseflies is best achieved by depriving them of breeding places. Although there is no way to control and remove all decaying organic matter, we have the power to control access to our food and cover and make inaccessible the garbage generated by humans.

The fly does have an enemy. *Eupusa muscae* is a relative of the common black mold found on bread (dead organic matter). Because *Eupusa muscae* feeds exclusively on living flies, it is a major agent in fly population control. Every year millions of flies worldwide become infected by the parasite and almost always perish before adding to the fly population.

But humans continue to try to control the flies that annoy them personally. The weapon of choice is the daily newspaper, a method less than 10 percent successful. The fly's ability to escape impending newspapers is not supernatural; it results from the fly's physical makeup. The hundreds of hairs that cover its body (and help it carry diseases hither and yon) are also its defense mechanism, for they are very sensitive to air pressure. Movements of a hand, a descending newspaper, or for that matter any solid object will create air fluctuation. The fly, warned of approaching trouble, escapes in less than an eye blink.

An effective weapon would need to have numerous holes through which air could freely pass. Such an ingenious invention, developed many years ago, is the flyswatter. It can descend on its victim with no air displacement, so the fly is cornered and squashed before it can figure out what has just happened. SWAT! One fly down—only 5,499,999,999,999 to go!

Strange Matrimonial Rituals

The romantic act of a groom carrying his bride over the threshold of her new home, still popular in many European countries, was a general custom in the United States during the 19th century. The new husband, proudly and with fanfare, would carry his wife through the doorway of the newly built cabin in which they would begin their lives together.

The custom had its origins in ancient Rome as a result of the Roman's belief in good and bad spirits. Everyone knew that any new house harbored two spirits, one good and one evil, that waited inside for the newly married couple to enter. The evil spirit was intensely jealous of human happiness and hoped to guarantee some misery by tripping the bride as she entered. That certainly would have dampened her good spirits, because tripping was a very harmful omen.

Any Roman would have had a terrible time sleeping at night if he or she were not also protected by a good spirit. It comes as no surprise that the bad spirit was always present on the person's left (sinister) side, and the good spirit, overflowing with virtue and "rectitude," occupied the right. The Romans had to be aware of which spirit was where; if a nervous bride stepped into the new house on her left foot, the evil spirit not only could trip her but could then enter the house any time it wished and make her life miserable. Ever-mindful of the left-sided spirit and to avoid such a mishap as tripping or stumbling, every Roman husband carried his new bride over the threshold. This completely bewildered the evil one, who could never again enter their home. Fortunately for the husband, brides were generally teenaged, lightweight, and easy to carry into the house without mishap.

This scenario could doubtless be rolled back much further in human history. Primitive man of the Stone Age probably dragged or carried his newly captured mate into his cave. This practice of abduction resurfaced among the Goths in northern Europe about A.D. 200. Although the men married women of their own community, capturing brides-to-be from another village was necessary whenever eligible women were in short supply.

Any young girl away from the safety of her home was fair game. With the help of a good companion (the "best man") the prospective groom would sweep the girl off her feet, and carry her, screaming and kicking, to his village. Following a hastily performed wedding ceremony, the groom would carry his bride over the threshold of their home. His best man continued to serve as a guard throughout the ceremony and remained outside their home in case the bride's family should attempt to steal her back. His usefulness waned as her compliance grew and as her family came to appreciate having one less person to share their meager food and space.

A bride will almost always wear a veil that is quite transparent so her beauty shows through. The custom has several possible origins. For example, it may be a relic of the canopy held over the couple's head in many ancient ceremonies. Such a canopy is still used today in Jewish weddings; it is supposed to protect the bride from the "evil eye" of anybody who might wish her ill, such as a jilted suitor.

The veil probably had its origin in the Muslim practice of purdah, which forbade a man to see his fiancee's face until the actual wedding. In accordance with the custom all unmarried women were secluded and kept entirely covered. After the ceremony the groom lifted the veil in anticipation of viewing a most beautiful sight for the first time; his bride's face. Thin screams of horror have been known to shoot from a groom's mouth when he beheld the face of his bride for the first time. This was a manner in which a desperate father married off a homely offspring. Naturally a most impressive dowry had to go with it.

The wearing of a veil had a definite purpose in the past and has continued as an important custom. When the bride glides toward the altar to take her wedding vows, her white veil is prominent; she will seem undressed without it.

Weird Remedies

New, tried and true, surefire remedies for headaches are not a recent phenomenon. They have been prescribed since the first pain in the head struck the first prehistoric human. The original remedy may have been the most sensible: crawl back in the cave away from noise, strain, and falling rocks and sleep it off.

The added suggestion to "take two aspirins" is probably the best quick fix. But purveyors of painkillers continue to proclaim theirs as the greatest breakthrough in the history of pain, and televised commercials demonstrate the effectiveness of each new remedy with a staged miraculous recovery. Sufferers seeking relief will purchase and try yet another panacea until it is replaced by a newer and better cure.

Before the days of pharmacies, some of the guaranteed remedies were quite bizarre. Among the preaspirin headache cures: rubbing cow dung on the temples or tying the head of a buzzard around the sufferer's neck. Anyone who prefers to actively participate in the therapy might lean the offending head against a tree while someone else drives a nail into the opposite side of the trunk. This remedy comes with the recommendation that the tree be thick enough that the nail will not be driven into the aching skull. If all of these remedies prove ineffective, one might try sleeping with a pain-cutting pair of scissors under the pillow. These were just a few of the many headache remedies used in various regions of the United States before the turn of the century.

Although such remedies were always passed on with a testimony of approval, no one seemed to know or care how they worked. Perhaps doing something, anything, was better than just enduring. And applying treatment after treatment may have provided therapeutic distraction.

Since headaches are one of the various agonies associated with a hangover, many civilizations have tried specialized remedies that can cure them all. The problem with hangover remedies is that they generally haven't worked, so new cures have to be invented with demanding regularity.

Ancient cures of hangovers were as numerous and as outlandish as those of modern times, and probably just as ineffective. A sure cure among ancient Romans was to douse themselves with cabbage, the rationale being that supreme deities dropped their seeds in cabbage and lettuce. Therefore, if people with a hangover ate or rubbed themselves with these vegetables, they would become godlike and their heads would stop aching. Pliny the Elder recommended downing two raw owl's eggs; this remedy may also have been effective in emptying the stomach. The Assyrians chronicle two of their most dependable cures: they would rub lemons in their armpits and, if that didn't work, would gulp a teaspoon of ground-up swallow's beaks! Within a day or two all hangover sufferers would experience relief and proclaim a new miracle of healing.

In more recent times the Basque people, notorious for their ability to drink with no adverse effects, concocted the morning-after broth *caldo borracho*, or "drunken soup." The name certainly suggests, however, that where there is a cure there may well be an affliction. The simple ingredients are olive oil, garlic, saffron, eggs, and breadcrumbs. The broth is so easy to prepare that hangover victims may even attend to themselves. It does bear a resemblance to dishwater but is rather tasty. And who knows? It may share the winner's circle with chicken soup for therapeutic qualities!

A Giant of a Kangaroo

The immediate mental image that will cross most minds when thinking of Australia is the kangaroo. Although the kangaroo is restricted to the Australian continent, it ranks among the most popular and recognizable animals in the world.

Kangaroos are quite varied in size, the smallest species being no larger than a jackrabbit. The largest, the red kangaroo, stands about seven feet, weighs 200 pounds, and has a head the size of a sheep's. With support from its powerful tail, it can jump 25 feet from a dead stop; when racing full speed, it can cover 40 feet in one leap.

Although impressive, the record of the red kangaroo pales next to its ancestor of the Pleistocene, which also lived on Australia, probably during one of the ice ages, when many animals reached their peak in size. The giant kangaroo of about 100,000 years ago stood 10 feet high and possessed a head as large as a Shetland pony's. With its enormous hind legs, feet, and tail, its running jump could have been almost 100 feet. This giant among kangaroos became extinct before the arrival of humans in Australia.

The World Herd

Of the many species of antelope, probably none is as hardy, efficient, and indestructible as the Arabian oryx. It is one of the more beautiful hoofed mammals, white or fawn in color with two long, scimitar-like horns so straight that in profile there appears to be but a single horn.

The Arabian oryx is a beast of formidable endurance. It withstands extremes in desert temperatures that may range from freezing to 140 sun-drenched degrees. It grazes on sparse tufts of grass and stunted shrubs, with occasional succulent bulbs and tubers to provide moisture. Because it feeds at night, when the plants have absorbed moisture, it can go many days without water. Its horns and hoofs are functional

both in digging for food and in fending off attacking lions, trucks, or tribesmen.

For centuries desert dwellers of the Near East have admired the oryx for its strength and toughness. Unfortunately, being admired is frequently an animal's undoing. The belief that the man who kills an oryx will inherit its courage and vigor was enough to fuel centuries of their slaughter as a certificate of manhood.

By the early 18th century the Arabian oryx was already regarded as a rarity. The shah of Persia preserved a small herd in his private park as curiosities to amaze and amuse his guests. At the beginning of the 20th century, the oryx was virtually wiped out everywhere except in the southern desert of Saudi Arabia known as Ar Rub' al-Khali, meaning "the empty quarter." One of the world's most inhospitable deserts, it lies scorched under the relentless glare of the sun, and the temperature rarely drops below 120 degrees at midday.

About 100 oryxes inhabited this desert at the turn of the century. A real test of manhood was to go hunting in Ar Rub' al-Khali, and Arab sheiks mounted on camels would go forth to demonstrate their virility by tracking the surviving herd. The contest was almost equally matched, because the chase was grueling and the oryxes were hard to find and quick to flee. Few kills were actually made, and the herd thrived. After World War II the discomfort and hardship factors were remedied as Arabian potentates found easier ways to attain the oryx's legendary strength and endurance. Armed with machine guns, these men in search of bravery set out in caravans of jeeps. As many as 300 men at a time would roar over the dunes, spraying a deadly hail of bullets at the panicking oryx herd.

So efficient were these massacres that by 1960 the last of the world's remaining oryxes seemed to have disappeared. Wildlife conservationists exploring Ar Rub' al-Khali could locate only two survivors, and one of those soon died of a bullet wound. The other, a female, was taken to the London Zoo. Later a herd of about two dozen was discovered in a remote desert area. In addition, the king of Saudi Arabia owned a small herd that was being bred to supply royal hunts. A British conservation group, the Fauna Preservation Society, realized that a few more machine-gun safaris would mean the end of the species, so they raised funds to collect a herd of Arabian oryxes that could be reared in captivity.

In 1962 a seven-man expedition went into the Arabian desert. Using spotter planes, they covered 6,000 miles of emptiness and located and caught only four oryxes. One of the four died after its capture. A high-caliber bullet was discovered in its leg; it too was a delayed victim of a hunting party.

The three healthy oryxes were shipped to the Phoenix Zoo in Arizona, where as expected they adapted easily in a climate similar to that of Arabia. They were joined shortly by the female from the London Zoo, and a fifth oryx was presented by the sultan of Kuwait. This became the nucleus of the "world herd" of Arabian oryxes. In October 1963 an oryx calf was born, and King Saud of Arabia was persuaded to part with four oryxes from his private game farm. By the end of that year the herd numbered 12.

Although a single herd is a fragile bulwark against total extinction, the oryxes at the Phoenix Zoo continued to thrive and multiply. To provide diversity in the genetic pool, several of the breeding stock were sent to zoos in Texas and California. In 1988 100 Arabian oryxes were reintroduced into a wild habitat in Oman, southeast of Saudi Arabia. An inventory taken in 1990 listed 494 registered individuals in Oman and 722 in zoo breeding programs.

The wild herd has been uncommonly successful, probably because it is protected by desert rangers. These rangers, 30 years before, may have been among those searching for the Arabian oryx to provide them with courage, strength, and endurance. Happily the oryx has helped them to discover the real qualities of manhood.

Hazards of Being a Bird

The basic problem of a bird's lifestyle is maintaining a temperature as high as 104 degrees Fahrenheit. Because birds are warm blooded (homeothermic), getting enough food is an urgent and ever-present need. Birds that spend much time in flight are relatively small and lose body heat more rapidly than heavier ones do. So they are notoriously voracious: compared with the elephant that eats 2 percent of its weight per day, the hog 4 percent, and the cold-blooded crocodile a mere 0.7 percent, a bird will eat up to 30 percent of its weight every day.

The law of surface-size-to-body-weight ratio has almost caught up with the hummingbird. This tiny bird is at the lowest limit of size at which a warm-blooded creature can exist. The hummingbird's heart takes as much space as it can afford and still leave room for other essential organs. Compared with the human heart, which is 0.42 percent of total body weight, the hummingbird's heart is 2.37 percent and beats at the awe-inspiring rate of 615 times per minute. The bird has learned one desperate expedient: it hibernates every night. In this condition of dormancy its body temperature goes down almost to that of the environment, and its need for oxygen dips correspondingly. If the hummingbird's ancestors had not evolved this function, it would starve to death overnight.

Several other species of birds can, when food is scarce, become temporarily cold-blooded and dormant. By reverting to the cold-bloodedness of their ancestral reptilians, the nestlings of the European swift can survive at least 10 days of complete fasting, an unusual ordeal for a young and growing bird. Another hibernator is the whippoorwill. Its body temperature drops from a normal 102 degrees to 65 degrees Fahrenheit, its breathing stops, and its digestive processes cease. This "false hibernation" can continue for several months until food is again available.

Most birds avoid the problems of winter cold by migrating southward. Those that do not must develop some means of maintaining warm-blooded body temperatures in the face of a rough winter. Freezing rain, for example, is as dangerous to birds wintering in the north as it is for bush pilots in their single-engine planes. During the night, freezing rain can fasten the wings of chickadees and juncos to their backs. At dawn when they try to fly, the birds fall helplessly to the snow and most likely become the next meal for a predacious hawk or eagle.

Some wintering birds such as the grouse and partridge are able to detect when a severe arctic storm is coming. They escape by flying full speed into a bank of soft snow, thus insulating themselves against the windchill and burying themselves without a telltale mark for predators. Occasionally they would be better off dealing with the swift death handed out by a predator, because their refuge may be iced over by freezing rain and the sanctuary becomes a tomb.

Birds that have evolved the ability to survive the hostile Arctic winter—discounting the fate of some icebound grouse, partridge or chickadee—face the hardship of finding adequate food in an environment where food is ever so scarce. But those that habitually migrate south for the winter do not find a bed of roses either. Migration is the great adventure for the majority of bird populations, but it also involves their greatest risks. Just why birds migrate is not completely explainable, but without doubt the choice to fly off results from transcendental memory (instinct) and not some individual or group decision.

Among the travelers from the northern hemisphere are the 15 species of wild geese that make formidable journeys southward each fall to their wintering ranges and return north as far as the Arctic to their breeding grounds. Each population follows a specific flyway hundreds or thousands of miles each way; one species crosses the Himalayas.

The wild goose does not waste precious energy reserves in elaborate courtship rituals. A pair of geese engage in a simple neck-stretching dance at the wintering grounds, honk in mutual acknowledgement, and mate for life. Considering that two compulsive, devoted parents are necessary to hatch, raise, and educate a brood of goslings, monogamy and fidelity are the only efficient and ecologically sound behaviors.

As they prepare to fly north to the breeding grounds, wild geese are unable to store fat reserves as do the smaller birds; extra weight would jeopardize their ability to become airborne. Therefore the goose must constantly be feeding (and emptying) itself during the flight home. Fortunately, the long hours of daylight as they follow spring northward and the plentiful supply of young, nutritious food make the journey adequately successful.

Back on the breeding grounds the work really begins, as the female lays and incubates a clutch of eggs. The male keeps her nourished, so that both will be able to nurture, train, and protect the goslings. To have them ready for the great flight southward in the fall takes the best efforts of two geese, working full time. Where family values are concerned, the wild geese have no equal.

Black as a Diamond

The term *bort* refers to minutely crystallized gray or black diamond masses that are not usable for individual crystals or industrial application. Bort is therefore crushed to powder for grinding and polishing purposes. There have been rare, well-documented exceptions. In 1927 a 33-carat piece of bort was found to contain at its heart a small red diamond of exceptional quality. Eventually a 5.05-carat gem was cut. In another case the bort was of such quality that it was actually cut into a most famous and unique gem. This is the celebrated black diamond called the black Orloff, a 67.5-carat stone cut from a 195-carat rough of Indian origin.

Of more recent origin is the beautiful stone known as the Amsterdam. Whereas the Orloff is more of a dark, gunmetal color and partly translucent, the Amsterdam is totally black and impervious to light. The stone arrived at the offices of Drukker and Sons in Amsterdam in 1972. It was in a parcel of mined bort destined to be crushed into diamond powder or broken up into smaller pieces for other industrial purposes. At that time the 55.85-carat rough would have been valued at no more than five to six dollars a carat. It's now worth considerably more.

When the cutter began to cleave the stone he immediately became aware both of its exceptional hardness and of the most important fact that the splinters were not in the least transparent but of the deepest black. So a wise decision was made to cut and polish the stone. The result is a pear-shaped, 145-facet black diamond weighing 33.74 carats, one of the rarest gems in the world. Unlike any other black diamond, it remains completely opaque, even when submitted to the strongest light.

Named in honor of the 700th anniversary of the city of Amsterdam, the diamond is in the possession of its discoverers, the Amsterdam diamond merchants D. Drukker and Sons.

National Drink of Ancient Egypt

Alcoholic beverages of various sorts have been causing high spirits and hangovers since the days of the Egyptian pharaohs. Barley beer was discovered around 4200 B.C. and rapidly became the national drink of Egypt. It was the cause of significant problems almost from the start.

Ancient hieroglyphics indicate that "houses of beer," the equivalent of modern saloons, began springing up about 2980 B.C. and that by 2160 B.C. alcohol abuse was common. The Egyptians even had their equivalent of the present-day happy hour. It was called "Day of Intoxication" to honor Hathor, the goddess of drinking.

Harsh measures were used to deal with those who overindulged in alcohol. Some wall paintings portray drunkards being flogged mercilessly with a stout stick and then imprisoned. These appear to be the lucky ones, as those in alcoholic stupor were often left on the banks of the Nile to be robbed, beaten, or even eaten by crocodiles.

Relaxed laws concerning public intoxication resulted in public orgies such as those in which Antony and Cleopatra are said to have indulged. Such parties, common for the upper class, compared in intensity to the uninhibited merriment of wild parties through the ages. One anonymous historian cites the case of a drunken Egyptian lady "who became immoderately mad, and afterward so lascivious that she immediately embraced every man she met. From laughing and singing she went over to rage and wanted to fight everybody."

Another Egyptian scribe describes a mother who delivered "three loaves of bread and two jars of beer daily" to her son (age unknown) who was away at school. Professors of that era often warned their students that "beer causes the soul to wander."

The ancient Egyptians are credited with inventing drunk tanks, steam rooms for hangovers, and drying-out shelters. Boozing it up became a boon for pharaohs; when the royal coffers were reaching the empty mark, they would quickly slap heavy taxes on beer.

Women in Early Medicine

In 14th-century Europe, women were forbidden to study at university levels and were refused licenses to practice medicine. They were permitted to continue working as midwives and "wise women," but the laws restricting them from other medical procedures were vigorously enforced.

In 1322 five women were brought to trial by the medical faculty at the University of Paris for practicing medicine without a license. One of the defendants was Jacqueline Felicie de Almania. The charges brought against her at the trial were that "she would cure her patients of internal illness and wounds. . . . She would visit the sick and examine their urine in the manner of physicians, feel the pulse and touch the body and limbs." She was quoted as having said to a sick person, "I will cure you by God's will, if you will trust in me." She made a pact with patients and received money from them.

The main witness against her was the surgeon John of Padua. Since women could not practice as lawyers, John declared, it was even more important that they not practice medicine, because losing a life was far more serious than losing a legal decision. Jacqueline was not without her own witnesses to testify in her behalf. Eight of de Almania's patients testified that she had been able to cure them when male doctors had failed to do so. The charges against her were not that she was incompetent but that as a woman she dared to cure at all.

De Almania also testified, arguing that women patients receive better treatment from a female physician, who would be able to examine their "breasts, belly, and feet" while the male counterpart could not. A male doctor was permitted to examine a female patient only while she was under a blanket or sheet and could diagnose only what he was able to feel. Even those in the courtroom at the time agreed with her testimony.

The judge remained unconvinced and stated, "It is certain that a man approved in the aforesaid art could cure the sick better than any woman." All five women were found guilty and received a most severe penalty for that historical era: they were all excommunicated from the Church.

Green Polar Bears

Repeated stories over the years of hunters and trappers spotting green polar bears were met with skepticism. They seemed to be an Inuit version of the pink elephants described so vividly by overindulged barflies. The stories persisted but went unbelieved; after all, light in Arctic regions can play strange tricks. Did not Coleridge's *Ancient Mariner* happen upon an iceberg described as "ice, mast high, came floating by, as green as emerald"?

Disbelief came to an abrupt halt in 1967, when Canadian scientists surprised themselves and everyone else by verifying the tales of green polar bears. A bear had been shot whose hair was definitely green. The unfortunate animal was sick when discovered, and a mercy killing was performed to spare it further suffering. When its hair was examined under a microscope, the scientists found that the hollow tube in the center of each follicle had been invaded by a form of parasitic green algae. It was a recent adaptation for this type of algae.

A bear thus invaded becomes quite disadvantaged when surrounded by an all-white environment. Unless, of course, the bear is careful to stand in front of an emerald green iceberg.

In Unity Is Strength

In most herd animal groups, newborn infants have very few minutes to get on their sticklike legs and be able to run with the herd. These are the most dangerous few minutes in the life of the newborn, for all sorts of predators such as hyenas, jackals, and lions prowl around the vicinity of the herd waiting for easy prey. Sometimes the herd will protect the mother and calf during and after the birthing. The following account was witnessed by several scientists, who recorded the incident on film.

A newborn wildebeest calf was having difficulty getting to its feet. Sensing a favorable opportunity, a hyena approached. Immediately a number of adults formed a broad front, protecting the calf still thrashing on the ground. Wildebeest beside wildebeest presented a formidable row of sharp horns, and the hyena skulked away. Almost immediately afterward a lioness attempted to approach the newborn. Normally the big cat's appearance would have resulted in the herd galloping off in wild flight. Instead they closed ranks, once again forming a menacing vanguard. This courageous defiance impressed the lioness, and she didn't attack the horned front line but slunk off to safer ground. By that time the calf had gained its feet and was able to run with the herd.

Naturalists have observed that wild geese, when numerous, will draw into a close battle line against a fox. In the face of this determined spirit, even an animal so passionately fond of geese (between its jaws) has no choice but to retreat.

The springboks in South Africa used a variation on the close defensive line. Toward the end of the 18th century they were so numerous that they loped across the plains in herds of up to 50,000 animals. According to reliable accounts, when an inexperienced young lion attempted to attack them they would encircle the predator and force it to march along in their midst. If the lion was unable to escape, it would be entrapped by the springbok herd until it collapsed from hunger or exhaustion and then would be trampled to death.

A rather remarkable event occurred near the town of Surat, India. A 15-year-old shepherd was attacked by a young tiger. The sheep immediately gathered into a dense formation and ran at the surprised tiger; it turned and fled in complete panic. The shepherd boy emerged from the encounter terrified but uninjured.

On the principle of strength in unity, starlings en masse are a formidable challenge to the hawk, which considers a single starling an easy feast. Ornithologists have often witnessed the bad judgment of a hawk attacking a flock (actually a murmuration) of starlings in a field. A thousand or more birds will instantly form an assault squadron that overwhelms the hawk as it circles the field and, in a mere second, forces it to land. Unable to take off again, the hawk become easy prey.

Not just animal enemies but even inanimate hazards are often surmounted by comrades of the herd, flock, or colony. Moreover, group support is provided not only by higher life forms but also by robotlike ant societies. In fact, if the others of the colony did not respond to their distress signals, the leaf-cutting ants of tropical regions might have become extinct some eons ago.

After heavy rainstorms, the tunnels in the underground labyrinths of leaf-cutting ants frequently collapse. The buried ants produce SOS signals, and rescue squads set off to find them and dig them out. By rubbing the rough surfaces of their abdominal segments together, the buried victims produce an ultrasonic noise picked up by vibration-sensitive cells on the feet of the search and rescue team. The liberators are able to trace the source of the distress calls directionally. Rescue includes reopening the collapsed passageway and getting the buried leaf-cutters back on the job.

Other defense mechanisms of leaf-cutting ants are even more elaborate. These ants are most vulnerable while cutting up leaves or carrying bits of leaf back to the fungus garden; since their mandibles are in use they are completely defenseless. On spring afternoons a tiny fly, appropriately called the "ant beheader," dive-bombs the worker and deposits an egg on the victim's neck. The fly larva hatches and bores into the ant's head, gradually eating until it is hollow. Then the larva bites off the hollowed-out head and uses it as a capsule in which to pupate. Such indiscriminate slaughter must be met head-on, and the leaf-cutters have an antiaircraft squadron to deal with the flies. Groups of dwarf ants that usually work underground in the fungus gardens accompany the parade of leaf-cutters. As soon as the flies attack, several dwarfs surround the leaf-cutter and snap at the enemy with their mandibles. When they have routed the threatening beheaders, they climb on the leaf the worker is carrying and taxi back to the nest. Even then they act as antiaircraft troops by keeping the ant beheaders out of their way.

Coffin for the Dinosaurs

The demise of the dinosaurs, so sudden from a geologic point of view, has been a subject of scientific debate for decades. Not only universities but B movies, comic strips, and dime novels have accepted the death of the dinosaurs as a sudden, mysterious event. Recent evidence has been unearthed that appears to provide convincing explanations.

The dinosaurs lived on earth during the 140 million years of the Mesozoic Era, the age of reptiles, divided into three periods: Triassic, Jurassic, and Cretaceous. A very successful and diverse family of animals, they ruled supreme, occupying every major niche on the land, with related species controlling the sea and air. The dinosaurs died out at the end of the Cretaceous period, 65 million years ago, and the age of reptiles was over.

The succeeding era, the age of mammals, began with the Tertiary Period. The boundary separating it from the earlier period is identified as the K-T boundary because it separates the Cretaceous (K) from the Tertiary (T). After 140 million years, the family of giants that ruled during the Mesozoic Era was suddenly gone.

What caused the extinction of such a successful group of animals? One of the many theories was overspecialization. Until two decades ago, textbooks proclaimed that dinosaurs became so specialized that they were unable to adapt to changing climates and fell victim to environmental modifications they couldn't handle. But the most persistent theories and opinions, even among scientists, were that the dinosaurs were too stupid to survive. This writer, supported by many authorities, once taught that if a student were to walk up to a brontosaurus and kick it in the tail, a full two minutes would pass before the animal, with its dull wits and even duller senses, would be aware of it.

In the 1980s evidence began to demonstrate that dinosaurs were not sluggish, but were far more agile and better adapted to their world than had been imagined. Some may even have been warm-blooded. Rather than the stupid brutes depicted earlier, many had quick reflexes and highly developed senses. At the end of the Cretaceous period, some 75 percent of the known species died out very suddenly on the land, sea, and air. Problem solvers turned to physical causes, such as a drastic fall in sea level, global cooling, and atomic radiation from space brought in by the explosion of a nearby star. Some scientists still prefer this last theory even though no hard evidence supports it.

The Cretaceous extinction was especially severe among sea creatures; in fact, they were more vulnerable than animals on land. Ammonites, so well known to fossil collectors as shelled, jet-propelled relatives of the octopus and squid, vanished from the sea along with sea monsters such as the pleisiosaurs, the mosasaurs, and the ichthyosaurs. The simultaneous termination of a wide variety of life on land and sea devalues any explanations peculiar to the dinosaurs. Even if the newly evolved caterpillars destroyed vegetation as thoroughly as do locusts, or some ratlike mammals developed a hearty appetite for dinosaur eggs, or a drastic disease struck, as several theories suggest, none would affect life beyond the beaches. Another recent theory about the dinosaur's demise targets the rise of the modern flowering plants. This involuntary change in diet deprived the dinosaurs of the laxative oils present in the older conifers, ferns, and cycads, so they all perished of constipation—no small matter for a dinosaur.

Investigators have found that teeming microspecimens of plankton, constituting the pastures of the sea, disappeared in a half-inch of sediment just at the end of the Cretaceous period; the event that ended the Cretaceous evidently was quick and catastrophic. This dismisses all hypotheses proposing a gradual change in climate, ecology, or even dinosaurian bowel habits.

During the late 1970s, Walter Alvarez, a geologist from the University of California, Berkeley, was conducting field research on late-Mesozoic rock layers exposed near the town of Gubbio, Italy. He and his associates were concerned with periodic changes in the earth's magnetic poles. Such paleomagnetism was recorded in the rocks from which he was collecting samples for laboratory research. Out of a brown-green clay that seemed to be advertising that it had something to tell them, Walter cut a sample and took it back to Berkeley to show to his father, Luis Alvarez, a physicist at Berkeley. Luis Alvarez suggested measuring the amount of iridium in the clay to find out whether the clay had been deposited over a short period or had accumulated gradually.

Iridium is an extremely rare element in the earth's outer crust. What little iridium could be measured in the clay layer would probably have come from meteorites, which along with comets are rich in iridium. The amount of iridium in any layer provides an estimate of how quickly the layer accumulated.

The result came as a shock: the amount of iridium in the Gubbio clay was 30 times the amount in the adjacent limestone. This clearly indicated that the earth was suddenly deluged with iridium. Even more startling results came from the same clay layer in other locales. One Danish specimen showed the amount of iridium increasing 160-fold, and a Belgian specimen reached levels 460 times normal.

After puzzling over these extraordinary results for several weeks, Luis Alvarez and his colleagues reached a most startling conclusion: the iridium might have come from a single, very large meteorite striking the earth. They estimated that an asteroid or a comet would have to be 6 to 12 miles in diameter to supply the almost 200,000 tons of iridium found in the clay layer.

Scientists know that the speed of an approaching asteroid would be about 150,000 miles per hour. As it ripped through the atmosphere and impacted the earth, it created tremendous pressures and temperatures, as much as 1 million times atmospheric pressure and several thousand degrees. The extreme violence of the collision completely vaporized the asteroid and about 10 times its mass of target rock, and the material sprayed outward from the expanding crater at 50 times the speed of sound. The blast, which released a thousand times the energy of the world's nuclear arsenals, punched a broad hole several miles deep in the crust and mantle. The fireball created by the impact expanded upward, carrying some 400 cubic miles of debris high above the atmosphere. (The 1980 eruption of Mount St. Helens ejected less than a third of a cubic mile of ash into the atmosphere.) Much of the finer material from the asteroid impact drifted in the stratosphere for years before it eventually settled to the ground, coating the entire globe with a thin layer of iridium-rich dust. The quantity of dust from an asteroid 10 miles in diameter would have blotted out the sun completely, and for a number of years there would have been unending night.

Without sunlight, plants would stop growing, and the ensuing famine would explain the disappearance among the fossil species. Microscopic plants in the sea died out almost completely and consigned to oblivion about 80 percent of marine life. The survivors were presumably those that could scavenge in the mud for the remains of plants of former years. The putrefying remains of plants and animals floating down the streams must have sustained some freshwater animal life, including the sole survivor of the great reptiles, the crocodile.

Blundering about in pitch darkness, the plant-eating dinosaurs stripped every last leaf from bushes and trees. The meat eaters made a regal banquet of their herbivorous cousins and then, enraged by hunger, turned cannibal. A few small animals, such as worms, insects, and birds, survived the long night by feeding on decaying vegetation, seeds, nuts, and one another. Among them were some of the diminutive mammals that lived mostly underground. When the sunlight finally broke through again, bathing the land with brilliant energy, these small mammals founded the new dynasties in the noncompetitive, depopulated world.

The Berkeley team published its hypothesis in 1980 but stressed that it was unproved by hard evidence. Many scientists opposed the impact catastrophe theory, probably because changes in the earth are expected to happen slowly. The first severe test for the idea was soon fulfilled successfully. If dust was scattered all around the world by the impact, then the extraterrestrial iridium should show up on the other side of the globe, as far away as possible from Europe and North Africa. In a short time samples were obtained from a site near Canterbury, New Zealand. To date well over 150 sites have been shown to contain the extraterrestrial-impact clay layer that marks the K-T boundary.

Evidence began to pile up, and by the early 1990s a collision of an enormous extraterrestrial body with the earth 65 million years ago was hard to refute. The location of the impact crater is now known, and the sequence of events set in motion by the colossal collision has become established.

What evidence supports the theory of an extraterrestrial collision? Some has been described in detail: the pencil-thin layer of meteoric iridium, found worldwide. One argument, that the excess iridium could have come from unusually massive volcanic outbursts, was quickly discarded. Further studies of the clay layer turned up more evidence, a form of mineral called "shocked quartz." Quartz crystals occasionally display a series of parallel lines on the face of the mineral as evidence of microscopic fractures. This occurs only when quartz is subjected to sudden extremely high pressures from an enormous explosion. This shocked quartz is found in debris of all major meteorite impact craters but not in volcanic debris.

Additional evidence common in the clay layer was tiny glassy spherules called "tektites," which are formed when molten droplets of melted rock shoot out from an explosion and resolidify into glassy beads. The abundance of shocked quartz and tektites occurring together in the iridium layers confirms that the K-T boundary included debris from a mighty explosion, consistent with an asteroid impact.

In the mid-1980s, scientists discovered that one of the components found in the layer all over the world was soot. To generate the amount of carbon in this soot layer, more forests would have to burn down than exist in the world today! Scientists conclude that most of the world's forests burned at the end of the Cretaceous, strong evidence of an event much different from assorted volcanic eruptions.

Some paleontologists believed that the dinosaurs had peaked and were on their way to extinction before the impact. In a 1989–91 field study of the Hell Creek Formation in the Upper Great Plains, an exceptional yield of late Cretaceous fossils showed that dinosaur extinction was synchronized with the impact. The study of 4,100 bones collected in the area indicated that ecological diversity had remained constant, with no evidence of decline in numbers or species.

Several groups of scientists studied the possible effects of a tsunami. They calculated that, if the impact had happened in an ocean, it would have created a splash more than three miles high near the impact site. Coastal areas adjacent to the site would have been inundated by a massive wall of water hundreds of feet high. In some regions the K-T boundary contained chaotic tumbled rocks, telltale deposits of a tsunami. The discovery of rocks of both continental and deep sea origin in the tsunami deposits provided evidence that the asteroid hit near a continental margin. Evidence of a tsunami was found on the perimeter of the Gulf of Mexico, following the shoreline that existed 65 million years ago. The absence of tsunami deposits elsewhere in the world helped to confirm a Central American impact.

The most awesome effect was proposed by an expert on cratering. Using computer models he calculated that much debris was thrown out of the atmosphere into space, only to fall back again. This created a concentrated rain of meteors streaking through the atmosphere for a few hours. It must have seemed that the sky was falling. The sky was ablaze with meteors, and the radiant energy level on the ground was like that inside an oven. Many above-ground animals must have been broiled to death where they stood. This apparently happened worldwide, and would explain why forests everywhere burst into flame.

Spectacular as the coastal devastation and fires may have been, each was a short-term effect. Surviving plants and animals encountered two longer-range problems. First, the shock of energy from the impact acted to convert large amounts of the atmosphere into nitrogen oxides, thus bringing on acid rain. In addition to other discomforts, survivors had to contend with the sort of pollution that has damaged forests in industrial North America and Europe, but at much higher concentrations. Many scientists believe that acid rain may have been a major agent in the extinctions.

The second long-term effect may have been more visually spectacular. Scientific teams have calculated that the enormous amount of debris ejected into the atmosphere included a high-altitude dust pall that fell back into the atmosphere within hours, dense clouds of dust carried away from the impact site by winds, and a global pall of smoke from the fires. Scientists estimate that about two pounds of dust was initially suspended above every square foot of Earth's surface. Within a week, Earth was densely clouded by dust and smoke; day turned to night.

The dust pall's most astonishing effect was to block sunlight. For one or two months, it was too dark to see; for two or three months, it was as dark as full moonlight; and for as much as a year, it was too dark for most plants to perform photosynthesis. The period of darkness destroyed the light-dependent plankton that float in the surface layers of the ocean, which would have had a rapid, drastic effect on marine life in general. The fossil record confirms that most life in the shallow seas was wiped out abruptly.

But where exactly was the smoking gun, the impact crater left by the deadly missile from outer space as it struck the earth? In 1978, Glen Penfield, a geologist with the Mexican national oil company, discovered a subterranean feature that had the characteristics of an impact crater. He pieced together material from magnetic surveys and gravity maps of the Yucatán peninsula and found a magnificent 100-mile bull's-eye hidden beneath the peninsula, with the town of Chicxulub, Mexico, at its center.

Dr. Alan Hildebrand of the University of Arizona, who was actively engaged in the search for the K-T impact crater, became aware of Penfield's find and joined forces with him. His search for more evidence led him to exposures of the K-T boundary 600 miles away, in southern Haiti. Green-brown clay impregnated with excess iridium, shocked quartz, and tektites was a signature of a giant asteroid, and coarse, jumbled rock fragments and debris scoured from one spot and dropped elsewhere were evidence of seismic sea waves hundreds of feet high.

News of Chicxulub spread to the Alvarez team of scientists, who searched in northeastern Mexico for more evidence of the crash. They discovered layers of undisturbed rock covered with deposits that were laid down violently, suggesting a monster wave. The impact scene was capped by beds of fine sediments of dust particles that were returning to earth, laced with iridium, after being lofted into the air by the impact. This was a recorded history of the area before, during, and after the giant asteroid had struck.

Radiometric dating of outcrops in Haiti and Mexico showed the age of 65 million years, the same as the well cores of Chicxulub; now enough evidence was gathered to re-create the impact. Within seconds after a space rock smashed to earth at 150,000 miles per hour, the crater, a hole 186 miles in diameter (the size of Connecticut), was blasted out to a depth of six miles. The energy of the impact created instant temperatures of over 20,000 degrees Fahrenheit, sent gigantic tidal waves roaring across the ocean, and triggered magnitude-12 earthquakes hundreds of miles away. Within hours dust began to form in the atmosphere, thickening into a cloud cover that completely enveloped the planet. Day became perennial night, and from space the veiled earth would have resembled the planet Venus.

So many ways to die! The life-forms of earth had many choices, all of them fatal. Scientists have recently added a new piece of evidence that spelled doom for the dinosaurs. The great agent of destruction was a global haze of sulfuric acid that blocked sunlight and plunged the planet into a dark, killing chill that lasted for decades. The new theory stems from recent geologic studies of the buried crater Chicxulub, in Mexico, which is believed to be the impact site. Finding the rock in the crater to be unusually rich in sulfur, scientists concluded that if the sulfur had not been added to the atmosphere, the dinosaurs might well have survived the impact that changed the course of evolution.

Scientists further estimated that the collision of an asteroid 6 to 12 miles in diameter would have vaporized much of the sulfur and spewed more than 100 billion tons into the air. This would have filled the air with sulfur dioxide, casting a pall that reeked of the devil's own brimstone, and a sulfuric acid haze in the upper atmosphere as the result of an interaction between sulfur and ultraviolet radiation. The release of sulfur caused by the impact would have been enough to plunge the planet into a perennial night lasting for decades and to shroud it in light-inhibiting sulfuric acid clouds.

The dust and soot from most of the debris would have drifted back to earth within six months, presumably too short a time for any global darkness to have caused mass extinctions. But the sulfur, being lighter, stayed aloft and created a dense and durable haze that covered the entire planet for several decades. According to most scientists, droplets of sulfuric acid would have filtered out sunlight by as much as 20 percent, producing a nuclear winter state 40 to 100 years long. Such a dramatic change in climate persisting over decades subjected organisms all over the world to long-term stresses. Few earth inhabitants could adapt in so short a period of time. The demise of the dinosaurs was probably complete in less than 100 years.

Such an enormous concentration of sulfur was present in less than 5 percent of the earth's crust. Had the asteroid struck almost any other place on the planet, it would not have generated the tremendous amount of sulfur spewed into the atmosphere.

Beneath that layer of iridium-laced clay the dinosaurs thrived; above it there were none. One can look on the clay layer as the lid on the "coffin for the dinosaurs."

BIBLIOGRAPHY

Books

Alpers, Antony. *Maori Myths and Tribal Legends.* Cambridge, MA: Houghton Mifflin, 1966.

Andrews, Roy Chapman. *Meet Your Ancestors.* New York: Viking, 1945.

Attenborough, David. *The Atlas of the Living World.* Boston: Houghton Mifflin, 1989.

———. *The First Eden.* Boston: Little, Brown, 1987.

———. *Life on Earth.* Boston: Little, Brown, 1987.

———. *The Living Planet.* Boston: Little, Brown, 1984.

Bahn, Paul G., and Jean Vertut. *Images of the Ice Age.* New York: Facts on File, 1988.

Baker, Robin, ed. *The Mystery of Migration.* New York: Viking, 1981.

Bakker, Robert T. *The Dinosaur Heresies.* New York: Morrow, 1986.

Ballantine, Bill. *Nobody Loves a Cockroach.* Boston: Little, Brown, 1968.

Bjagat, Shantilal P. *Creation in Crisis.* Elgin, IL: Brethren, 1990.

Blaney, Walter M. *How Insects Live*. New York: Elsevier-Phaidon, 1976.

Blond, Georges. *Great Migrations*. London: Hutchinson, 1958.

Breasted, James Henry. *Ancient Times*. New York: Ginn, 1935.

Bronowski, J. *The Ascent of Man*. Boston: Little, Brown, 1974.

Brown, Bruce, and Lane Morgan. *The Miracle Planet*. New York: Gallery Books, 1990.

Burton, Maurice. *Animal Legends*. New York: Coward-McCann, 1957.

Burton, Maurice, and Robert Burton. *Encyclopedia of Fish*. New York: Finsburg Books, 1984.

———. *Inside the Animal World*. New York: Quadrangle, 1977.

Caird, Rod. *Ape Man: The Story of Human Evolution*. New York: Macmillan, 1994.

Calder, Nigel. *Timescale*. New York: Viking, 1983.

———. *The Comet Is Coming*. New York: Viking, 1981.

Callahan, Philip S. *Insects and How They Function*. New York: Holiday House, 1971.

Caras, Roger. *Dangerous to Man*. New York: Stoeger, 1974.

Carr, Donald E. *The Deadly Feast of Life*. New York: Doubleday, 1971.

Carrington, Richard. *Mermaids and Mastodons*. New York: Rinehart, 1957.

Carswell, John, and others. *Splendors of the Past*. Washington, DC: National Geographic Society, 1981.

Ceram, C. W. *The First American*. New York: Mentor, 1971.

———. *Hands on the Past*. New York: Knopf, 1966.

———. *Gods, Graves and Scholars*. New York: Knopf, 1959.

Chard, Chester S. *Man in Prehistory*. New York: McGraw-Hill, 1969.

Chinery, Michael. *Killers of the Wild*. New York: Chartwell Books, 1979.

Chorlton, Windsor. *Ice Ages*. Alexandria, VA: Time-Life Books, 1983.

Clark, Eugenie. *The Lady and the Sharks*. New York: Harper & Row, 1969.

Clausen, Lucy W. *Insect Fact and Folklore*. New York: Macmillan, 1954.

Coe, William R. *Tikal: A Handbook of the Ancient Maya Ruins*. Philadelphia: University of Pennsylvania, 1988.

Colbert, Edwin H. *Dinosaurs: An Illustrated History*. Maplewood, NJ: Hammond, 1983.

———. *Evolution of the Vertebrates.* New York: Wiley, 1980.

———. *Wandering Lands and Animals.* New York: Dutton, 1973.

———. *Men and Dinosaurs.* New York: Dutton, 1968.

Constable, George. *The Neanderthals.* New York: Time-Life Books, 1972.

Couper, Heather, with Nigel Henbest. *New Worlds: In Search of the Planets.* Reading, MA: Addison-Wesley, 1986.

Cousteau, Jacques-Yves, with James Dugan. *The Living Sea.* New York: Harper & Row, 1963.

Cousteau, Jacques-Yves, and Philippe Cousteau. *The Shark: Splendid Savage of the Sea.* New York: A & W Visual Library, 1970.

Daniels, George G., ed. *Volcano.* New York: Time-Life Books, 1982.

de Blij, H. J., and others. *Nature on the Rampage.* Washington, DC: Smithsonian Books, 1994.

Dennis, Jerry. *It's Raining Frogs and Fishes.* New York: Harper Collins, 1992.

Desmond, Adrian J. *The Hot Blooded Dinosaurs.* New York: Dial, 1976.

Dickinsen, Joan Younger. *The Book of Diamonds.* New York: Crown, 1965.

Ditmars, Raymond L. *Thrills of a Naturalist's Quest.* New York: Macmillan, 1932.

Douglas-Hamilton, Iain, and Oria Douglas-Hamilton. *Among the Elephants.* New York: Viking, 1975.

Dröscher, Vitus B. *They Love and Kill.* New York: Dutton, 1976.

———. *The Friendly Beast.* New York: Dutton, 1971.

Dryson, James L. *The World of Ice.* New York: Knopf, 1962.

Edey, Martland. *The Missing Link.* New York: Time-Life Books, 1972.

Ehrlich, Paul, and Anne Ehrlich. *Extinction.* New York: Random House, 1981.

Erickson, Jon. *Ice Ages: Past and Future.* Blue Ridge Summit, PA: Tab Books, 1990.

Fabre, J. Henri. *The Life of the Spider.* New York: Dodd, Mead, 1919.

Fagan, Brian. *The Journey from Eden.* London: Thames & Hudson, 1990.

———. *The Great Journey.* London: Thames & Hudson, 1987.

———. *The Rape of the Nile.* New York: Scribner's, 1975.

Fox, Robin Lane. *Alexander the Great.* London: Dial, 1973.

Frazier, Kendrich. *The Violent Face of Nature*. New York: Morrow, 1979.

Frisch, Karl von. *Animal Architecture*. New York: Harcourt Brace Jovanovich, 1974.

———. *Ten Little Housemates*. New York: Pergamon, 1960.

Fuchs, Sir Vivian, ed. *Forces of Nature*. New York: Holt, Rinehart & Winston, 1977.

Fuller, Errol. *Extinct Birds*. New York: Facts on File, 1987.

Gallenkamp, Charles. *Maya: The Riddle and Rediscovery of a Lost Civilization*. New York: David McKay, 1976.

Gardom, Tim, with Angela Milner. *The Book of Dinosaurs*. Rocklin, CA: Prima, 1993.

George, Jean Craighead. *Beastly Inventions*. New York: David McKay, 1970.

George, Uwe. *In the Deserts of This Earth*. New York: Harcourt Brace Jovanovich, 1976.

Goodall, Hugo, and Jan van Lawich. *Innocent Killers*. New York: Ballantine Books, 1970.

Gore, Al. *Earth in the Balance*. Boston: Houghton Mifflin, 1992.

Gould, Stephen Jay. *Hen's Teeth and Horse's Toes*. New York: Norton, 1983.

Gowlett, John. *Ascent to Civilization*. New York: Knopf, 1984.

Grant, Michael. *The Rise of the Greeks*. New York: Scribner's, 1987.

Griffin, Donald R. *Animal Minds*. Chicago: University of Chicago Press, 1992.

Groves, Don. *The Oceans*. New York: Wiley, 1989.

Grzimek, J. C. Bernhard, ed. *Animal Life Encyclopedia*. Vols. 1–13. New York: Van Nostrand Reinhold, 1984.

Hadingham, Evan. *Secrets of the Ice Age*. New York: Walker, 1979.

Hallam, A. *A Revolution in Earth Science*. New York: Clarendon, 1973.

Hartmann, William K., and Ron Miller. *The History of Earth*. New York: Workman, 1991.

Hayes, Harold T. P. *The Last Place on Earth*. New York: Stein & Day, 1977.

Heilbrin, Angelo. *Mont Pelée and the Tragedy of Martinique*. Philadelphia: Lippincott, 1905.

Helm, Thomas. *Hurricanes: Weather at Its Worst*. New York: Dodd, Mead, 1967.

Hoffman, Michael A. *Egypt Before the Pharaohs*. New York: Knopf, 1979.

Hohn, Reinhardt. *Curiosities of the Plant Kingdom*. New York: Universe Books, 1980.

Hoke, John. *Discovering the World of the Three-Toed Sloth*. New York: Franklin Watts, 1976.

Honders, John, ed. *The World of Birds*. New York: Peebles, 1975.

Horner, John R., and James Gorman. *Digging Dinosaurs*. New York: Workman, 1988.

Horner, John R., and Don Lessem. *The Complete T. Rex*. New York: Simon & Schuster, 1993.

Imbrie, John, and Katherine Palmer Imbrie. *Ice Ages: Solving the Mystery*. Cambridge, MA: Harvard University Press, 1986.

James, Peter, and Nick Thorpe. *Ancient Inventions*. New York, Ballantine, 1994.

Jenkins, Alan C. *Mysteries of Nature*. New York: Facts on File, 1984.

Jenkinson, Michael. *Beasts Beyond the Fire*. New York: Dutton, 1980.

Johanson, Donald, and Maitland Edey. *Lucy*. New York: Simon & Schuster, 1981.

Johanson, Donald, Lenora Johanson, and Blake Edgar. *Ancestors: In Search of Human Origins*. New York: Random House, 1994.

Johanson, Donald, and Kevin O'Farrell. *Journey from the Dawn*. New York: Viking, 1990.

Johanson, Donald, and James Shreeve. *Lucy's Child*. New York: Morrow, 1989.

Krupp, E. C. *Echoes of the Ancient Skies*. New York: Harper & Row, 1983.

Lane, Frank W. *The Violent Earth*. Topsfield, MA: Salem House, 1986.

———. *The Elements Rage*. New York: Chilton, 1965.

Lavine, Sigmund. *Strange Travelers*. Boston: Little, Brown, 1960.

Leakey, Mary. *Disclosing the Past*. New York: Doubleday, 1984.

Leroi-Gourhan, Andre. *Treasures of Prehistoric Art*. New York: Harry Abrams, 1967.

Lessem, Don, and Donald F. Glut. *Dinosaur Encyclopedia*. New York: Random House, 1993.

Lewin, Roger. *Thread of Life*. Washington, DC: Smithsonian Books, 1982.

Ley, Willy. *Exotic Zoology*. New York: Bonanza Books, 1987.

———. *The Poles*. New York: Time, 1962.

Lineaweaver, Thomas H. III, and Richard H. Backus. *The Natural History of Sharks*. New York: Lyons & Burford, 1984.

Lockley, Martin. *Tracking Dinosaurs*. Cambridge: Cambridge University Press, 1991.

Lopez, Barry Holstun. *Of Wolves and Men*. New York: Scribner's, 1978.

MacGregor, Alasdair Alpin. *The Changing Land*. Bath, England: Kingsmead, 1973.

Mackal, Roy P. *Searching for Hidden Animals*. New York: Doubleday, 1980.

Maeterlinck, Maurice. *Life of the Bee*. New York: Dodd, Mead, 1936.

Maffei, Paolo. *Monsters in the Sky*. Cambridge, MA: MIT Press, 1976.

Man, John. *The Day of the Dinosaur*. London: Bison Books, 1978.

Martin, Richard Mark. *Mammals of the Oceans*. New York: Putnam, 1977.

May, John, and Michael Marten. *The Book of Beasts*. New York: Viking, 1982.

McCormick, Harold W., Tom Allen, and William Young. *Shadows in the Seas*. New York: Weathervane Books, 1968.

McGowan, Christopher. *Dinosaurs, Spitfires, and Sea Dragons*. Cambridge, MA: Harvard University Press, 1991.

McInerny, Derek, and Geoffrey Gerard. *All About Tropical Fish*. New York: Macmillan, 1966.

McIntyre, Loren. *Amazonia*. San Francisco: Sierra Club Books, 1991.

McLoughlin, John C. *Archosauria: A New Look at the Old Dinosaur*. New York: Viking, 1979.

McNeely, Jeffrey A., and Paul Spencer Wachtel. *Soul of the Tiger*. New York: Doubleday, 1988.

McNeill, William H. *Plagues and Peoples*. New York: Doubleday, 1976.

Minton, Sherman A., and Madge R. Minton. *Venomous Reptiles*. New York: Scribner's, 1969.

Moody, Richard. *A Natural History of the Dinosaurs*. London: Hamlyn, 1977.

Moore, Patrick. *Armchair Astronomy*. New York: Norton, 1984.

Morrison, Philip, and Phylis Morrison. *The Ring of Truth*. New York: Random House, 1987.

Motz, Lloyd, ed. *Rediscovery of the Earth*. New York: Van Nostrand Reinhold, 1975.

Myles, Douglas. *The Great Waves*. New York: McGraw-Hill, 1985.

Norman, David. *Dinosaur!* New York: Prentice-Hall, 1991.

O'Neil, Paul. *Gemstones*. New York: Time-Life Books, 1983.

Panati, Charles. *Extraordinary Origins of Everyday Things*. New York: Harper & Row, 1987.

Paul, Gregory S. *Predatory Dinosaurs of the World*. New York: Simon & Schuster, 1988.

Platt, I. *A Library of Wonders and Curiosities*. New York: Alden, 1884.

Preiss, Byron, ed. *The Universe*. New York: Bantam, 1987.

Priess, Byron, and Robert Silverberg, eds. *The Ultimate Dinosaur*. New York, Bantam, 1992.

Preston, Douglas, Jr. *Dinosaur in the Attic*. New York: St. Martin's Press, 1980.

Psihoyos, Louis, with John Knoebber. *Hunting Dinosaurs*. New York: Random House, 1994.

Quammen, David. *The Flight of the Iguana*. New York: Delacorte, 1988.

Raup, David M. *Extinction: Bad Genes or Bad Luck?* New York: Norton, 1992.

Reader, John. *Missing Links*. Boston: Little, Brown, 1981.

Ricciuti, Edward. *Killer Animals*. New York: Walker, 1976.

———. *Killers of the Sea*. New York: Macmillan, 1973.

Ritchie, Carson I. A. *Insects: The Creeping Conquerors*. New York: Elsevier/Nelson, 1979.

Ritchie, David. *The Ring of Fire*. New York: Mentor, 1982.

Robinson, Victor. *The Story of Medicine*. New York: New Home Library, 1943.

Ruiz, Fernando Medina. *Mayan Culture*. Mexico, D.F.: Panorama Editorial, 1988.

Sarton, George. *A History of Science*. Cambridge, MA: Harvard University Press, 1953.

Scheffer, Victor. *The Year of the Whale*. New York: Scribner's, 1969.

Schultz, Gwen. *Ice Age Lost*. New York: Doubleday, 1974.

Schweighauser, Charles A. *Astronomy from A to Z*. Springfield, IL: Sangamon State University, 1991.

Sears, Paul B. *Deserts on the March*. Washington, DC: Island Press, 1935.

Shapiro, Harry L. *Peking Man*. New York: Simon & Schuster, 1974.

Sheehan, Angela, ed. *The Prehistoric World*. New York: Warwich, 1975.

Sherratt, Andrew, ed. *The Cambridge Encyclopedia of Archeology*. New York: Crown, 1980.

Short, Nicholas M. *Planetary Geology*. Englewood Cliffs, NJ: Prentice-Hall, 1975.

Shreeve, James. *The Neandertal Enigma*. New York: Morrow, 1995.

———. *Nature: The Other Earthlings*. New York: Macmillan, 1987.

Sieveking, Ann. *The Cave Artists*. London: Thames & Hudson, 1979.

Silverberg, Robert. *Mammoths, Mastodons and Man*. New York: McGraw-Hill, 1970.

———. *The Auk, the Dodo, and the Oryx*. New York: Crowell, 1967.

———. *The Morning of Mankind*. New York: Graphic Society, 1967.

Smith, Howard E. *Killer Weather*. New York: Dodd, Mead, 1982.

Smithsonian Institution. *Fire of Life*. New York: Norton, 1981.

Sparks, John. *The Discovery of Animal Behavior*. Boston: Little, Brown, 1982.

Spaulding, David A. E. *Dinosaur Hunters*. Rocklin, CA: Prima, 1993.

Stevens, Payson R., and Kevin W. Kelley. *Embracing Earth*. San Francisco: Chronicle Books, 1992.

St. John, Jeffrey. *Noble Metals*. Alexandria, VA: Time-Life Books, 1984.

Stivens, Dal. *The Incredible Egg*. New York: Weybright & Talley, 1974.

Stommel, Henry, and Elizabeth Stommel. *Volcano Weather*. Newport, RI: Seven Seas, 1983.

Stringer, Christopher, and Clive Gamble. *In Search of the Neanderthals*. New York: Thames & Hudson, 1993.

Sullivan, Walter. *Landprints*. New York: New York Times Books, 1984.

———. *Continents in Motion*. New York: McGraw-Hill, 1974.

Sunset Books, *National Parks of the West*. Menlo Park, CA: Lane, 1980.

Sutcliffe, Jenny, and Nancy Duin. *A History of Medicine*. New York: Barnes & Noble, 1992.

Sutton, Ann, and Myron Sutton. *Wildlife of the Forests*. New York: Harry Abrams, 1979.

———. *Nature on the Rampage*. New York: Lippincott, 1962.

Tarbuck, Joseph J., and Frederick K. Lutgens. *The Earth*. Columbus, OH: Merrill, 1984.

Tattersall, Ian. *The Human Odyssey*. New York: Prentice-Hall, 1993.

Teale, Edwin Way. *Strange Lives of Familiar Insects*. New York: Dodd, Mead, 1962.

———. *The Insect World of J. Henri Fabre*. New York: Dodd, Mead, 1949.

Ternes, Alan, ed. *Ants, Indians and Little Dinosaurs*. New York: Scribner's, 1975.

Thomson, Peggy. *Auks, Rocks and the Odd Dinosaur*. New York: Crowell, 1985.

Thorndike, Joseph J., ed. *Mysteries of the Deep*. New York: American Heritage, 1980.

Vine, Louis L. *Your Dog: His Health and Happiness*. New York: Winchester, 1975.

———. *Dogs, Devils & Demons*. New York: Exposition, 1971.

Voyage Through the Universe: Stars. Alexandria, VA: Time-Life Books, 1989.

Walker, Charles. *Wonders of the Ancient World*. New York: Crescent, 1980.

Wallace, Joseph. *The Rise and Fall of the Dinosaur*. New York: Gallery Books, 1987.

Waltham, Tony. *Catastrophe: The Violent Earth*. New York: Crown, 1978.

Weiner, Jonathan. *Planet Earth*. New York: Bantam, 1986.

Wellnhofer, Peter. *The Illustrated Encyclopedia of Pterosaurs*. New York: Crescent, 1991.

Wendt, Herbert. *Out of Noah's Ark*. Boston: Houghton Mifflin, 1959.

Weyer, Edward M., Jr. *Strangest Creatures on Earth*. New York: Sheridan House, 1953.

Whipple, A. B. C. *Storm*. Alexandria, VA: Time-Life Books, 1982.

Wickler, Wolfgang. *Mimicry in Plants and Animals*. London: World University, 1968.

Wilford, John Noble. *The Riddle of the Dinosaur*. New York: Knopf, 1985.

Willis, Delta. *The Hominid Gang*. New York: Viking, 1989.

Worth, C. Brooke. *Mosquito Safari: A Naturalist in Southern Africa*. New York: Simon & Schuster, 1971.

Wylie, Francis E. *Tides*. Brattleboro, VT: Stephen Greene, 1979.

Young, Louise B. *The Blue Planet*. Boston: Little, Brown, 1983.

———. *Earth's Aura*. New York: Knopf, 1977.

Periodicals

Alper, Joe. "Shaking Seattle." *Earth*. July 1993.

Alvarez, Luis. "Mass Extinctions Caused by Large Bolide Impacts." *Physics Today*. July 1987.

Alvarez, Luis W., Walter Alvarez, and others. "Extraterrestrial Cause for the Cretaceous-Tertiary Extinction." *Science*. June 6, 1980.

Alvarez, Walter, and others. "Impact Theory of Mass Extinctions and the Invertebrate Fossil Record." *Science*. March 16, 1984.

──────. "Evidence for a Major Meteorite Impact on the Earth 34 Million Years Ago." *Science*. May 21, 1982.

"Analysis Suggests T. Rex Was Big Eater." *Redlands Daily Facts*. July 10, 1994.

Anderson, Robert B. "Bear Baiting." *Natural History*. December 1992.

Angier, Natalie. "Giant Fungus May Be Oldest Living Organism." *San Bernardino Sun*. April 2, 1992.

Archer, Michael. "World Furry-Weight Champions." *Natural History*. April 1994.

Associated Press. "Arkansas Gets a Buzz out of Mosquito Cookoff." *Redlands Daily Facts*. August 23, 1994.

"Back Home on the Range." *Environment*. November 1991.

Barklow, Williams. "Big Talkers." *Wildlife Conservation*. January–February 1994.

Bartusiak, Marcia. "Secrets of the Glaciers." *Discover*. January 1981.

Beatty, J. Kelly. "Killer Crater in the Yucatan." *Sky & Telescope*. July 1991.

"Beautiful, Dangerous Tourist Draw." *Telegraph*. (Nashua, NH). January 2, 1995.

Benford, Gregory. "Species on Ice." *Earthwatch*. July–August 1993.

Bishop, Ellen Morris. "The Mystery of the Owl." *Earth*. January 1993.

Bower, Bruce. "Tooth Analysis May Decipher Prehistoric Diets." *Science News*. October 20, 1990.

──────. "Rivers in the Sand." *Science News*. August 26, 1989.

──────. "Ancient Human Ancestors Got All Fired Up." *Science News*. December 10, 1988.

Boxer, Sarah, ed. "One, Two, Three Strikes You're Dead in This Old Ball Game." *Discover*. June 1986.

———. "Fossil Find May Be Earliest Known Hominid." *Science News*. April 14, 1984.

Brough, Holly B. "A New Lay of the Land." *World Watch*. January–February 1991.

"Celebration of a Volcano." *Science News*. January 24, 1987.

Clifton, Merritt. "Who's a Birdbrain?" *Animals' Agenda*. December 1990.

Clinton, Patrick. "The Wild File." *Outside*. May 1995.

Cook, William J. "The Invasion of Mars." *U.S. News and World Report*. August 23, 1993.

Cowen, Robert C. "Disappearance of Dinosaurs Offers Environmental Lessons." *Christian Science Monitor*. May 19, 1987.

"Cuba's Presence May Have Been Result of Asteroid, Scientists Say." *Redlands Daily Facts*. May 20, 1990.

Culotta, Elizabeth. "The Dinosaurs' Path to Extinction." *Earth*. January 1993.

Diamond, Jared. "Bob Dylan and Moas' Ghosts." *Natural History*. October 1990.

"Dino-Death: Flash Broil or Slow Steam." *Science News*. February 3, 1990.

Dodson, Peter. "Life Styles of the Huge and Famous." *Natural History*. December 1991.

Editorial Staff. "The Comeback of the Arabian Oryx." *Arizoo* (Phoenix Zoo) January–February 1988.

Esteve, Harry. "Earthquake Rattles Oregon." *Register Guard* Eugene, OR. March 26, 1993.

"Evolving Views of Dinosaurs." *Natural History*. December 1987.

Flanagan, Ruth. "Freddy Krueger of the Cretaceous." *Earth*. January 1993.

Freedman, David H. "Bolts from the Blue." *Discover*. December 1990.

Garrett, Wilbur E., ed. "La Ruta Maya." *National Geographic*. October 1989.

———. "Where Did We Come From?" *National Geographic*. October 1988.

Gould, James L. "Do Honeybees Know What They Are Doing?" *Natural History*. June–July 1979.

Gould, Stephen Jay. "An Asteroid to Die For." *Discover*. October 1989.

———. "The Lesson of the Dinosaurs." *Discover*. March 1987.

Grace, Beth. "Live Bacteria Discovered in Mastodon." *San Bernardino Sun*. May 4, 1991.

Grieve, Richard A. F. "Impact Cratering on the Earth." *Scientific American*. April 1990.

Hansson, Lennart. "The Lemming Phenomenon." *Natural History*. December 1989.

Hildebrand, Alan R., and William V. Boynton. "Cretaceous Ground Zero." *Natural History*. June 1991.

Hoke, John. "Oh, It's So Nice to Have a Sloth Around the House." *Smithsonian*. April 1987.

Holtz, Robert Lee. "Sulfur Clouds May Have Killed Dinosaurs." *Los Angeles Times*. December 30, 1994.

———. "Scientists Recover DNA from Time of Dinosaurs." *Los Angeles Times*. June 10, 1993.

Idyll, Clarence P. "Grunion, the Fish That Spawns on Land." *National Geographic*. May 1969.

Jastrow, Robert. "The Dinosaur Massacre: A Double-barreled Mystery." *Science Digest*. September 1983.

Kerr, Richard A. "End of the Dinosaurs." *Los Angeles Times*. June 12, 1989.

Kidd, Joe. "Quake Hit at Heart of Town." *Register Guard*. Eugene, OR. April 1, 1993.

Lang, John S. "White Bears, Black Gold." *Defenders*. January–February, 1992.

Letourneau, Deborah K. "Ants That Pay the Piper." *Natural History*. October 1993.

Lillywhite, Harvey B. "Sauropods and Gravity." *Natural History*. December 1991.

Matheny, Ray T. "An Early Maya Metropolis Uncovered—El Mirador." *National Geographic*. September 1987.

Matthews, Downs. "Don't Mess with Mom." *Wildlife Conservation*. March–April 1991.

McKean, Kevin. "Dinosaur Dynasty." *Modern Maturity*. October–November 1986.

Mitchell, John. "The Universe in a Grain of Sand." *Earthwatch*. May–June 1993.

Monastersky, Richard. "Viking Teeth Recount, Sad Greenland Tale." *Science News*. November 12, 1994.

———. "Giant Crater Linked to Mass Extinction." *Science News*. August 15, 1992.

———. "Venus' Skin Shows Familiar Tectonic Scars." *Science News*. August 8, 1992.

———. "Counting the Dead." *Science News*. February 1, 1992.
———. "Closing In on the Killer." *Science News*. January 25, 1992.
———. "Dinosaurs' Swan Song: Out with a Bang." *Science News*. November 9, 1991.
———. "Reining In a Galloping Triceratops." *Science News*. October 20, 1990.
———. "Rattling the Northwest." *Science News*. February 17, 1990.
———. "Reopening Old Wounds." *Science News*. January 20, 1990.
———. "Signs of an Ancient Worldwide Wallop." *Science News*. March 4, 1989.
Morell, Virginia. "How Lethal Was the K-T Impact?" *Science*. September 17, 1993.
Myers, Norman. "The Big Squeeze." *Earthwatch*. November–December 1993.
Nicholson, Thomas D. "Total Eclipse." *Natural History*. February 1979.
Novacek, Michael J. "A Pocketful of Fossils." *Natural History*. April 1994.
Ostrom, John H. "A New Look at Dinosaurs." *National Geographic*. August 1978.
Pendick, Daniel. "A Year in the Life of Mars." *Science News*. August 14, 1993.
Prospero, Joseph M. "Dust from the Sahara." *Natural History*. May 1979.
Quammen, David. "Voice Part for a Duet." *Outside*. August 1995.
———. "Phobia and Philia." *Outside*. September 1994.
———. "Everything Old Will Be New." *Outside*. January 1992.
Ryan, John C. "When Nature Loses Its Cool." *World Watch*. September–October 1992.
Sagan, Carl. "A Warning for Us?" *Parade Magazine*. June 5, 1994.
"Seal Weigh-In." *Discover*. July 1991.
Sereno, Paul C. "Dinosaurs and Drifting Continents." *Natural History*. January 1995.
Sharpton, Virgin L., and others. "Chicxulub Multiring Impact Basin." *Science*. September 17, 1993.
Sukumar, Raman. "Elephant Raiders and Rogues." *Natural History*. July 1995.
Svitil, Kathy A. "Hurricane from Hell." *Discover*. April 1995.
Tickell, Sir Crispin. "Hell and High Water." *Earthwatch*. September 1991.

Turbak, Gary. "Where the Buffalo Roam." *Wildlife Conservation*. December 1995.

Tuttle, Russell H. "The Pitted Pattern of Laetoli Feet." *Natural History*. March 1990.

Waters, Tom. "Ascent of the Bugs (and Flowers)." *Earth*. January 1994.

———. "The Dinosaur Acid Test." *Discover*. February 1990.

Whittow, G. Causey. "Night Shift for Sloths and Other Sluggards." *Smithsonian*. January 1977.

"Why Ice Floats." Signs of the Times. November 1980.

Wilder, Rachel. "When Laziness Pays Off." *Science Digest*. July 1983.

Wilford, John Noble. "Dinosaur Theory: Sulfur Was Villain (but Hero for Humans)." *New York Times*. January 3, 1995.

Wilson, Edward O. "Empire of the Ants." *Discover*. March 1990.

Wright, Karen. "Dinosaur Doctors." *Discover*. November 1991.

Wuethrich, Bernice. "Cascadia Countdown." *Earth*. October 1995.

INDEX

Achernar, Mount, 262
Acid rain
   and extinction of dinosaurs, 305
   on Venus, 153
Addo Elephant National Park (South Africa), 206
Adrian V, Pope, 44–45
Advertising, oldest, 19
Aepyornis maximus, 57
Africa, insect consumption in, 195–96
African elephants, 206–7
Ahuitzotl, 266
Air pressure, elevation and, 255–56
Alaska earthquake (1964), 159
Albucasis, 210
Alcohol, in ancient Egypt, 296
Alexander the Great, 197–99
Alexandria, Queen, 72
Alfonso XIII, King, 72
Alligators, gender selection in, 74
Alpha Centauri, 275
Alps, Hannibal's crossing of, 9
Altitude, and air pressure, 255–56
Alvarado, Pedro de, 184
Alvarez, Luis, 301, 302, 306
Alvarez, Walter, 301

Alveoli, 257
Amber, 176–77
American Revolutionary War, 114–15
Amethyst Cliff (Yellowstone National Park), 23–24
Ammonia, 176
Ammonites, 301
Amsterdam diamond, 295
Andes, human physiology in, 255–57
Anne, Queen, 16
Antarctica, 120
   drill cores of ice from, 177–78
   ice cap in, 121
   tropical climate in, 262–63
   whiteouts in, 175
Antelopes, 110–11, 116
Anthropology, 242
   coal, discovery of, 82
   footprints, fossilized human, 168–74
   mammoth meat, preserved, 66
   musical instruments, earliest, 154–55
   pit barbecues, 104
   stone throwing, 58–61
Antibiotics, 110
Antoinette, Marie, 76
Ants
   human consumption of, 196

   leaf-cutting, 299
   and termites, 226
Apatosaurs, 93
Apollo 12, 100
Ar Rub' al-Khali, 291–92
Arabian oryx, 290–93
Arachnids
   scorpion, 24–27
   trap-door spider, 186–87
Archaeology
   advertisement, oldest, 19
   Alexander the Great, 197–99
   Aztec conquest, 264–68
   cosmetics, 141–43
   Hannibal's elephants, 8–10
   Jewish holy oil, 77–78
   Mayans, 183–85
   "pig soldiers," 61
   sit-down strike, first, 279–81
   water travel, 40–41
Archosauria, 248
Argonauts, 122–23
Aristarchus of Samos, 138–40
Aristotle, 138
Arizona
   copper mining in, 84–86
   sandstorm in, 187
   Verde Lake Beds in, 260
Armillaria bulbosa, 67
Artworks, prehistoric, 169–71

Asian flu, 73
Assyrians, ancient, 25
Asteroid belt, 89
Asteroids
  and extinction of dinosaurs, 302–7
  near misses by, 90–91
Astronomy
  asteroids, 90–91
  Mars, 236–39
  meteorites, 88–90
  solar system, origin of, 197
  telescope, "time travel" with, 274–76
  Venus, 151–54
Atmosphere, composition of early, 176–77
Atmospheric pressure, 255
Atomic bomb, 38
Atomic energy, 37
Australopithecus afarensis, 60
Aztec, 264–68

Baboon, 233
Bacteriology, 108–10
  housefly, bacteria on, 285–86
  oldest living things, 67–68
Bald eagles, 234–35
Barbers, 209–11
Baychinco (phantom ship), 161–62
Bears
  earthquake forecasting by, 282
  polar, 78–82, 297
  prehistoric hunting for, 242
Beaulieu, Jacques, 16
Bedbugs, 53
Bee-eater (bird), 167–68
Beer
  in ancient Egypt, 296
  consumption of, by elephants, 117–18
Bees
  undertaker honeybee, 86–88
  as weapons, in warfare, 253
Bendire, Maj. Charles, 136
Bends, the, 130–31
Bernhardt, Sarah, 17
Best man, 288
Birds. *See* Ornithology
Bison, 192–93
Black diamonds, 295
Black mamba, 239–40
Black Orloff, 295
Blanchard, Jean-Pierre, 28
Blizzards, 140–41
Blood, of Andean residents, 257
Body hair, 264

Body temperature, of birds, 293–94
Boiga irregularis, 162–66
Bolivia, 202–3
Bombay, 25–26
Bort, 295
Botany
  cottonwood tree, 166
  "living fossils," 231–32
  paleobotany, 145–50
Bourne, Lizzie, 225–26
Breyfogle, Charles, 245–48
Brown tree snake, 162–66
Bruno, Giordano, 275
Butterflies, as jewelry, 258
Buzzard, turkey, 6–8

Caesar, 61
Caine Mutiny, The (Herman Wouk), 32
Caldo borracho, 289–90
California
  earthquakes in, 55, 157–58
  fossilized human footprints in, 168
  Great Mouse War in, 106–8
  sharks off coast of, 65
Calving, 121
Cannibalism
  among Maori, 97
  crocodiles, 251
  dance fly, 51
  king cobra, 43
  praying mantis, 49–50
  scorpion, 26–27
Carbon, in diamond, 58, 75
Carbon dioxide, in early atmosphere, 176, 178
Carboniferous Period, 148–50, 176
Carcharodon megalodon, 65
Carnotite, 38–40
Carpets, 25
Carthage, 8–10
Cascadia subduction zone, 159–60
Catnip, 116–17
Cave of the Witches, 171–72
Cecropia ants, 181
Celot, Jacques, 76
Centipedes, 149
Centrosaurus, 4
Ceratopsian dinosaurs, 129
Chapman, Ray, 60
Chicxulub crater, 306–7
Chilean earthquake (1960), 159
China
  coal seams in, 150
  coal use in, 82
  earthquakes in, 91–92, 283–84
Churchill (Manitoba), 81
Cinnabar, 143
Cleopatra, 142, 143

Climate, Martian, 238–39. *See also* Meteorology
Cline, Isaac, 269
Coal, 82, 147–50
Cobra
  king, 41–44
  and snake charmers, 205
Cobra (typhoon), 31–32
Cockroaches, 148
Cod, 10–11
Coelacanth, 229–31
Colchis, 123
Coleridge, Samuel Taylor, 297
Colonial America
  Great Plague and settlement in, 144
  and Little Ice Age, 114–15
  quackery in, 16
Colorado, uranium rush in, 40
Columbus, Christopher, 277–78
Commonwealth Bay (Antarctica), 120
Cones (of eye), 124
Connecticut, meteorite impacts in, 89–90
Continens Liber (Rhazes), 131–32
Continental drift, Antarctica and, 263
Cook, Capt. James, 93, 94
Copeina arnoldi, 12
Copernicus, Nicolaus, 139–40, 275
Copper Queen Library, 84–86
Cortez, Hernando, 184, 268
Corvidae, 167
Cosmetics, 141–43
Costa Rica, 258
Cottonwood tree, 166
Cowpox, 227
Creosote bush, 67
Cretaceous Period
  atmosphere during, 176, 177
  changes in, 150
  crocodiles during, 183
  mass extinction at end of, 300–301
  Tyrannosaurus rex in, 2
Crocodiles, 248–51
  and Cretaceous mass extinction, 303
  and dinosaurs, 221–22
  largest, 182–83
Crocodylus porosus, 182–83
Cro-Magnons, 105–6, 112, 141, 242
Crows, 167
Cuckoo, 200–202
Cyclones, 212

Dance flies, 51
Darkness, and predatory hunting, 110–11
Darwin, Erasmus, 34
De Almania, Jacqueline Felicie, 297
De Revolutionibus (Nicolaus Copernicus), 140
Death Valley, 245–48
Debtors, Indian, 204
Defoe, Daniel, 137
Deir el-Medina, 280–81
Desert Protection Act, 244–45
Devonian Period, 148
Diamonds
  black, 295
  burning, 58
  Hope Diamond, 75–77
  in ostriches, 261
Diana (hurricane), 214
Diego de Landa, Bishop, 185
Dinornis maximus, 95
Dinosaurs, 147
  and crocodiles, 183, 221–22
  egg of, petrified, 48–49
  extinction of, 300–307
  fossilized footprints of, 258–59
  gender selection in, 74–75
  herding by, 93
  injuries to, 128–31
  Maiasaura, 191–92
  Malerisaurus, 221–22
  Oviraptor, 188
  and oxygen in atmosphere, 177
  sauropods, 222–23
  stegosaurus, 39
  Therizinosaurus, 224–25
  Tyrannosaurus rex, 1–6, 223–24
Diodorus, 282
Dogs
  in first parachute, 28
  sense of smell in, 124–25
Douglas, Dr. James, 85, 86
Doyle, Arthur Conan, 228
Drachenhohle Cave, 242
Dragonflies, 148
Drones, 88
Duck-billed platypus, 220–21
Durango (Mexico), 25
Dust storms, 187–89

Eagles, bald, 234–35
Earth, roundness of, 137–40
Earthquakes, 157–60
  animal forecasting of, 282–84
  Gansu Province (China), 91–92
  legendary, 202–3
  sensing of, by animals, 55
Edict of Tours, 210

Edison, Thomas, 17
Edmontosaurus, 224
Eggs
  bedbug, 53
  crocodiles, 250
  cuckoo, 201
  fish, 11–14
  gender selection in alligator, 74
  honeybee, 86–87
  housefly, 286
  king cobra, 43–44
  loligo squid, 52
  moa, 96
  mouthbreeder, 214–15
  oology, 135–37
  petrified dinosaur, 48–49
  wild geese, 295
Egyptian plover, 251
Egyptians, ancient, 138
  advertising by, 19
  and beer, 296
  cosmetics use by, 141–43
  and crocodiles, 248–49
  and gold, 243–44
  sails, use of, 41
  sit-down strike by, 279–81
Elephant seals, 207–9
Elephants, 260
  African, 206–7
  intoxication of, 117–18
  king cobra attacks on, 42
  and "pig soldiers," 61
  in Punic Wars, 8–10
Ellsworth Base (Antarctica), 175
England
  Great Plague in, 143–44
  quackery in, 16
  water travel in Stone Age, 40
Enterobacter cloacae, 68
Entomology, 235
  bedbugs, 53
  dance flies, 51
  edible insects, 195–97
  fleas, 273–74
  housefly, 284–87
  jewelry, insects as, 258
  moths, 185–86
  praying mantis, 49–50
  scorpions, 24–27
  termites, 226–27
  trap-door spider, 186–87
  undertaker honeybee, 86–88
  wasps, 186–87
Epidemic, flu, 69–74
Eric the Red, 113
Ethiopia, 195–96
Eupusa muscae, 286
European swift, 293
Eurypterids, 24
Evolution, time scale of, 147
Extinction, Cretaceous, 300–301

Eye
  Egyptian reverence for, 142
  human vs. dog, 124

Fingernails, painted, 143
Fingerprints, 125, 264
Firestorms, 45–48
First Punic War, 8
Fishes. *See* Ichthyology
Fleas, 273–74
Fleming, Alexander, 110
Flood myths, 112–13
Flowering plants, 147, 150, 301
Flu epidemic (1918–1919), 69–74
Fluorescence, in scorpions, 26
Food and Drug Act of 1906, 17–18, 253
Food poisoning, in World War II, 252–53
Footprints, fossilized, 168–74, 258–61
Forest fires, firestorms in, 46
Forests
  first, 148–50
  fossilized, 23–24
Fossa, 55
Fossils
  footprints, 168–74, 258–61
  forests, fossilized, 23–24
  insect, 235
  living "fossils," 228–32
  and paleobotany, 145–47, 149
  and paleopathology, 127–31
  radioactive, 39
  shark teeth, 65
  trees, 262–63
  Triceratops, 6
  Tyrannosaurus rex, 3, 6
France
  dinosaur remains in, 48
  Niaux cave in, 170–71
Frankenstein's monster, 34–35
Frankincense, 141
Franklin, Benjamin, 101
Freud, Sigmund, 151
Frogs, mating ritual in, 50
Frost wedging, 122
Fulgurites, 101–2
Fungi, 109–10
  giant, 67

Galileo, 151, 274, 275
Galveston (Texas), 268–72
Gandidier, Alfred, 57
Gansu Province (China), 91–92
Gates, Sir Thomas, 278, 279
Gauls, 10
Geese, wild, 294–95, 298

Geology. *See also*
    Earthquakes;
    Mineralogy; Volcanoes
  of Antarctica, 263
  atmosphere, composition
    of early, 176
  ice ages, 111–16
  roundness of earth,
    137–40
  sandstorms, 187–89
George V, King, 72
Germany, firestorms in, 46
Giant kangaroo, 290
Giant moa, 58, 93–97
Giant sloth, 106
Gift-giving, by animals, 53
Ginkgo tree, 232
Gizzard, 95–96, 261
Glaciers, 115–16, 121
Glitter, 142
Global warming, 121, 178
Glossopteris, 262–63
Gobi Desert, 188
Gold, 243–48
  in seawater, 119
Golems, 33–34
Gondwanaland, 93
Gorillas, 193–94
  body hair on, 264
Goths, 288
Great Chicago Fire, 48
Great Fleet, 94, 96
Great Mouse War, 106–8
Great Plague, 143–44
Great white shark, 64, 65
Greeks, ancient, 138
  Alexander the Great and,
    197–98
  and idea of human
    perfection, 141–42
  Xerxes the Great and, 252
Green-backed herons, 167
Greenhouse effect, on Venus,
    152
Greenland
  colonists in, 113
  drill cores of ice from,
    177–78
  ice cap in, 121
  polar bears in, 81
Gregory X, Pope, 44–45
Grey, Sir George, 97
Ground sloth, 88
Grouse, 294
Grunion, 12–14
Guam, 162–66
Guatemala, 184
Gubbio (Italy), 301–2

Hadacol, 18
Hadrosaurs, 128–29
Hair, body, 264
Hairy Ape, The (Eugene
  O'Neill), 264
Halsey, Adm. William F.,
    31–33
Hamburg (Germany), 46

Hamilcar Barca, 9
Hangover cures, 289–90
Hanna, Wilson C., 136
Hannibal, 8–10
Haplochromis, 214–15
Harlan, Dr. Richard, 218
Hasdrubal, 8–10
Hawaii, shark legends in,
    102–4
Hawks, 299
  zone-tailed, 136
Headache remedies,
    288–90
Hearst, George, 248
Hell Creek Formation, 304
Hemingway, Ernest, 92
Henry I, 253
Herding
  and defensive actions,
    298–99
  by dinosaurs, 93
Hermes (asteroid), 90
Herodotus, 55–56, 143
Herons, green-backed, 167
Herpetology. *See* Crocodiles;
    Snakes
Hilara sartor, 51
Hildebrand, Dr. Alan, 306
Hindus, ancient, 138
Hippocrates, 70
Hitler, Adolf, 254
Hoatzin, 166–67
Holyoke (Massachusetts), 93
Homer, 188
Homo erectus, 172
Homoloichos and
  Anaxidamos, legend of,
    123–24
Honeybees, undertaker,
    86–88
Hong Kong flu, 73
Hope, Henry Thomas, 76
Hope Diamond, 75–77
Horner, John, 192
Horse latitudes, 134–35
Hottentots, 196
Housefly, 284–87
Huitzilopochtli, 266
Hull (destroyer), 32
Human sacrifice, by Aztec,
    266–67
Hummingbird, 293
Hunting
  for bears, 242
  by lions, 260
  for moas, 96–97
  by Neanderthal, 104–5
Hurricanes, 211–14
  Galveston (1900), 268–72
  Matecumbe Keys
    (Florida), 92
  and Shakespeare's
    Tempest, 277–79
  typhoons vs., 29
Hydrarchus, 219–20
Hydraulic cycle, 118–19
Hydrology, 120–22

Hyenas, 298
Hypselosaurus, 48

Ibex, 104–5
Icarus (asteroid), 90
Ice, 121–22
Ice ages, 121, 178
Icebergs, 121
Ichthyology
  coelacanth, 229–31
  land-spawning fish, 11–14
  lungfish, 231
  mouthbreeder, 214–15
  sharks, 62–66, 102–4, 199
  stingrays, 240–41
India
  beer-drinking elephants
    in, 117–18
  bees in, 253
  debtors in, 204
  mystics in, 203–4
  snake charmers in, 205
Indigirka River, 66
Influenza, 69–74
Initiation ceremonies, 170
Innocent V, Pope, 44–45
Insects. *See* Entomology
Intelligence
  of king cobra, 43
  in monkeys and apes,
    232–34
Intoxication, among animals,
    116–18
Inuit, 80–81, 112, 113, 195
Iridium, 301–4

Jaffe, Rabbi, 34
Jamestown, 113
Japan
  firestorm in, 45–46
  Mongol invasion of,
    29–30
Jason and the Golden Fleece,
    122–23
Jefferson, Thomas, 114
Jenner, Edward, 227–28
Jericho (ancient city), 59–60
Jerusalem, temple of, 77
Jewish folklore, 33–34
John of Padua, 297
John XXI, Pope, 45
Jourdain, Silvester, 278
Juan de Fuca Plate, 158–59
Judendorff, Gen. von, 71
Jurassic Park (film), 2
Jurassic Period, 39, 263

Kamikaze pilots, 30–31
Kangaroos, 290
Kanitovski, Prince, 76–77
Kant, Immanuel, 282
King cobra, 41–44
King crab, 229
King Kong (film), 1
Koch, Albert, 217–20
Kohl, 142
Krait, 41

K-T boundary, 300, 304–6
Kublai Khan, 29, 56

Laetoli, 172–73
Latimer, Marjorie, 229–31
Latin America, entomophagy in, 196
Le Tuc d'Audoubert (cave), 169–70
Leaf-cutting ants, 299
Legends
   Baychinco (phantom ship), 161–62
   Copper Queen Library, 84–86
   flood, 112–13
   headache remedies, 288–90
   Homoloichos and Anaxidamos, 123–24
   Jason and the Golden Fleece, 122–23
   matrimonial rituals, 287–88
   monsters, 33–35
   Robinson Crusoe, 137
   roc, 55–57
   shark, 102–4
   war stories, 252–55
Lemmings, 20–23
Leuresthes tenuis, 12–14
Lianoning Province earthquake (1975), 283–84
Lice, 195
Life, origin of, 147–48
Light, speed of, 275
Lightning, 99–102
   on Venus, 152–53
Limulus, 229
Lions
   elephants, attacks on, 260
   humans, attacks on, 132–34
   and springboks, 298–99
Lippershey, Hans, 274
Little Ice Age, 114
"Living fossils," 228–32
Lloyd George, David, 72
Locoism, 117
Locoweed, 117
Locusts, 195, 282
Loess, 187
Loligo squid, 51–53
London flu, 73
Louis XIV, King, 75, 76
Louis XV, King, 76
Louis XVI, King, 76
Lowe, Rabbi Judah, 34
Lowell, Percival, 236
Lungfish, 231
Lungs, 257

Macedonia, 197–98
Madagascar, 55–57
Magellan (spacecraft), 153
Maggots, 196

Maiasaura, 191–92
Malachite, 142
Malaysia, 258
Malerisaurus, 221–22
Mamba, 41, 239–40
Mammoths, 66, 105–6
Manganese, 119
Manna, 195
Maori, 94, 96, 97
Marco Polo, 56, 82
Marie Louise, Queen, 76
Marine volcanoes, 10–11
Mariner (space probe), 236
Mars, 153, 236–39
Mars Observer (space probe), 236–37
Masai, 132–33
Mastodons, 68, 218–20, 259–61
Matecumbe Keys (Florida), 92
Mateer, Dr. Warren D., 136
Mather, Cotton, 227
Mating and reproduction. *See also* Eggs
   bacteria, 108–9
   bald eagle, 234–35
   crocodiles, 250
   elephant seals, 208–9
   fish, 11–14
   housefly, 286
   king cobra, 43–44
   lemmings, 21
   moa, 96
   rituals, mating, 49–54
   scorpion, 26–27
   sloth, 180
Matrimonial rituals, 287–88
Max, Prince of Baden, 72
Mayans, 110, 184–85
McLean, Evelyn Walsh, 77
Medea, 123
Medicine. *See also* Pathology
   antibiotics, 110
   and barbers, 209–11
   flu epidemic, 69–74
   headache remedies, 288–90
   medieval female physicians, 296–97
   paleopathology, 127–31
   Petrus Hispanus, 44–45
   quackery, 15–19
   smallpox vaccine, 227–28
Megatherium, 88
Melville (research ship), 10–11
Mesozoic Era, 93, 221, 300. *See also* Cretaceous Period
   climatic changes during, 74–75
   mild climate during, 115
Metellus, Lucius Caecitius, 8
Meteor Crater (Arizona), 90

Meteorites, 88–90
   and extinction of dinosaurs, 302–7
Meteorology. *See also* Hurricanes
   blizzards, 140–41
   firestorms, 45–48
   ice ages, 111–16
   lightning, 99–102
   Mount Washington, 225–26
   sandstorms, 187–89
   tornadoes, 83
   typhoons, 29–33
   whiteouts, 174–75
   windiest place in the world, 120
Mice, in Great Mouse War, 106–8
Michigan, Peshtigo Horror firestorm in, 46–48
"Middle of nowhere," 15
Migration, bird, 294
Mineralogy
   black diamonds, 295
   diamonds, 75–77
   diamonds, burning, 58
   gold, 243–48
   oceanic minerals, 118–20
   uranium rush, 37–40
Mississippian Period, 235
Missourium, 217–20
Moa, giant, 58, 93–97
Mohammed, 141
Monaghan (destroyer), 32
Mongols, 29–30
Monsters, legendary, 33–35
Montana, *Maiasaura* nests in, 191–92
Montego, Francisco de, 184
Montezuma II, 265, 268
Mosasaurs, 129–31
Moslems, 25
Mosquitoes, 196
Moths, and fire, 185–86
Mountain sickness, 256
Mountebanks, 16–17
Mouthbreeder, 214–15
Mummies, 243–44
Musical instruments, earliest, 154–55
Myrrh, 141

Nairobi and Voi National Park, 133–34
Namib Desert, 25–26
NASA, 99–100
Native Americans, fossilized footprints of, 168
Natural History of Man, The (J. G. Wood), 59
Neanderthals
   barbecuing by, 104–5
   bear hunting by, 242
   cave dwellings of, 111–12, 171–72

cosmetics use by, 141
  in desert environments, 274
Nefertiti, Queen, 143
Nests, bald eagle, 234
New South Wales, 64
New York City, flu epidemic in, 72
New Zealand, giant moa in, 93–97
Niaux (cave), 170–71
Nicaragua, 168
Nile crocodile, 250, 251
Nile River, 41
1989FC (asteroid), 90–91
Nitrogen, 176
Norsemen, 113
North American Plate, 159
Northern terns, 53–54
Norway, lemmings in, 20
Noseprints, 125, 264
Nuclear power, 37–38
Nurse bees, 87

Oceanography. *See also* Pacific Ocean
  horse latitudes, 134–35
  minerals, oceanic, 118–20
Oil, Jewish holy, 77–78
Oldest living things, 67–68
Olduvai Gorge, 60
Oligocene epoch, 23
Olympus Mons, 238
O'Neill, Eugene, 264
Onesimus, 227
Onions, therapeutic value of, 19
Oology, 135–37
Ophiophagus hannah, 41–44
Ordovician Period, 229
Oregon
  earthquakes in, 158
  firestorm in, 46
Ornithology
  bald eagle, 234–35
  body temperature, regulation of, 293–95
  cuckoo, 200–202
  giant moa, 93–97
  hurricanes, birds in, 213
  intelligence of birds, 166–68
  northern terns, 53–54
  oology, 135–37
  ostriches, diamonds in, 261
  raven, 205
  robin, 27
  roc, 55–57
  turkey buzzard, 6–8
Oryx, Arabian, 290–93
Ostriches
  diamonds in, 261
  intoxicated, 116
Oviraptor, 188
Owen, Sir Richard, 94–95, 219

Oxygen
  in early atmosphere, 176, 177
  in water, 122

Pacific Ocean
  "middle of nowhere" in, 15
  typhoons in, 29–33
Paleobotany, 145–50
Paleomagnetism, 301
Paleontology. *See also* Dinosaurs
  footprints, fossil, 258–61
  forests, fossilized, 23–24
  Missourium hoax, 217–20
  paleopathology, 127–31
Paleopathology, 127–31
Paleozoic Era, 109
Panaceas, 18
Pangaea, 263
Parachuting, 28
Parallax shifts, 139
Parasuchus, 221–22
Paré, Ambroise, 211
Paricutín (volcano), 45
Partridge, 294
Pathology
  flu epidemic (1918–1919), 69–74
  Great Plague, 143–44
  smallpox, 227–28
Pearl Harbor, 103, 104
Penfield, Glen, 306
Penicillin, 110
Pennsylvanian Period, 235
Perfumes, 141
Perkins, Dr. Elisha, 16
Permian Period, 262, 263
Perouse, Count de la, 59
Pershing, Gen. John, 71
Peshtigo Horror, 46–48
Peter the Great, 150–51
Petrification, 39
  dinosaur egg, 48–49
Petrus Hispanus, 44–45
Pets
  scorpions as, 27
  sloths as, 181
Phelps-Dodge and Company, 85
Philip II, King of Macedon, 197, 198
Phobosuchus, 183, 248
Phosphate, 120
Phosphorite, 120
Pickford, Mary, 72
Pigs, as "soldiers," in ancient Rome, 61
Pioneer Venus (spacecraft), 152
Piper, Stanley, 107–8
Pit barbecues, 104–6
Pitcairn Island, 15
Pittsburgh (cruiser), 71
Plankton, 305–6
Plants. *See* Botany

Plate tectonics, on Venus, 153
Platypus, 220–21
Pleistocene Epoch, 116
Pliny the Elder, 289
Plutarch, 123–24
Pocahontas, 240
Poisonous animals
  Boiga irregularis, 162–66
  king cobra, 41–44
  scorpion, 24–27
Polar bears, 78–82
  green, 297
Pompeii, 254
Porpoises, 63
Port Royal (Jamaica), 202
Powhatan, Chief, 240
Praying mantis, 49–50
Psychology
  Peter the Great, 150–51
  Rhazes, 131–32
Ptolemy, 139
Punic Wars, 8–10
Purdah, 288
Pure Food and Drug Act of 1906, 17–18, 253
Puritans, 144
Pyramids, 183, 279
Pyrrhus, 61
Pythagoras, 138

Quackery, medical, 15–19
Quartz, shocked, 304
Queen Shark (legend), 104
Quetzalcoatl, 267–68

Radioactive fossils, 39
Raffia palm, 56
Ramses III, 281
Rao (Hindu yogi), 203–4
Rats, 282
Rattlesnakes, 282
Ravens, 205
Red blood cells, 257
Red kangaroo, 290
Red ocher, 141
Red snow, 189
Religion, Aztec, 265–68
Reproduction. *See* Mating and reproduction
Reptiles. *See* Crocodiles; Snakes
Revolutionary War, American, 114–15
Rhazes, 15–16, 131–32
Rhesus monkey, 232–33
Rhynchocephalian (order), 228–29
Richard Coeur de Lion, 253
Richter scale, 157
Rivers, 121
Robin, 27
Robinson Crusoe (fictional character), 137
Roc, 55–57
Rocks and minerals. *See* Mineralogy

Rods (of eye), 124
Romans, ancient, 8–10
  cosmetic use by, 142
  hangover cure of, 289
  and Jewish holy oil, 77
  and legend of Homoloichos and Anaxidamos, 123–24
  matrimonial rituals of, 287
  "pig soldiers" of, 61
Roosevelt, Franklin D., 72, 135
Roosevelt, Theodore, 135
Rouge, 143
Rule, Dr. John, 94
Ryan, Nolan, 58–59

Saber-toothed cats, 259–61
Sadiman (volcano), 172, 173
Sahara Desert, 188, 189
Sails, earliest, 41
St. Helens, Mount, 192, 303
Saint-Hilaire, Geoffrey, 57
Salt, 119–20
Samoa, 59
San Andreas Fault, 158
Sandstorms, 187–89
Sap, fossilized, 176–77
Sapsucker, 116
Sauropods, 222–23
Saw-scaled viper, 41
Schiaparelli, Giovanni, 236
Scipio Africanus, 10
Scorpions, 24–27
Sea snakes, 41
Seals, elephant, 207–9
Sears, Paul, 187
Second Punic War, 8–10
Seismic activity, animal reactions to, 55
Seismosaurus, 223
Selkirk, Alexander, 137
Shakespeare, William, 185, 278–79
Shark Research Panel, 63
Sharks, 199
  legends of, 102–4
  myths about, 62–66
  teeth of, 129
Sheep, 299
Shelley, Mary, 35
Shelley, Percy Bysshe, 111
Shem, 33–34
Sherman, Gen. William, 252
Shocked quartz, 304
Shooting stars, 89
Silurian period, 24
Sinbad the Sailor, 56
Sirocco, 189
Sit-down strike, first, 281–82
Skydiving, 28
Sloths
  giant, 106
  ground, 88
  three-toed, 178–82
Smallpox vaccine, 227–28

Smell, sense of, in dogs, 124–25
Smith, James L. B., 229–31
Smith, Capt. John, 240–42
Smithsonian Institution, 77, 136
Snake charmers, 205
Snakes
  black mamba, 239–40
  Boiga irregularis, 162–66
  king cobra, 41–44
  laziness of, 145
  rattlesnakes, 282
Snow
  in blizzards, 140–41
  red, 189
Snow blindness, 174
Solar system, 197
Songs, bird, 27
Sousa, John Philip, 17
Spain, flu epidemic in, 71
"Spanish Lady," 71
Spanish-American War, 253
Sphenodon, 228–29
Spider, trap-door, 186–87
Spielberg, Steven, 2
Spores, oldest bacterial, 67
Springboks, 298–99
Squid, loligo, 51–53
Starlings, 299
Stars
  light from, 275
  parallax shifts in, 139
Stegosaurus, 39
Stingrays, 240–41
Stone throwing, 58–61
Storms. See also Hurricanes
  blizzards, 140–41
  firestorms, 45–48
  lightning, 99–102
  sandstorms, 187–89
  typhoons, 29–33
Strachey, William, 278–79
Sulfuric acid, 307
Sulla (Roman general), 123–24
Swift, European, 293
Swine flu, 73

Taft (California), 106–7
Taipan, 41
Tasman, Abel, 94
Tavernier, Jean Baptiste, 75–76
Teeth
  decay, tooth, 109
  fossilized shark, 65
  of *Triceratops*, 5
Tektites, 304
Telescope, "time travel" with, 274–76
Temperature
  on Mount Washington, 225
  and oxygen absorption by water, 122

Tempest, The (William Shakespeare), 278–79
Tennant, Smithson, 58
Tenrec, 55
Termites, 196, 226–27
Terns, northern, 53–54
Tertiary Period, 300
Tezcatlipoca, 267
Therizinosaurus, 224–25
Thermometers, 114
Thomson, Maurice, 135
Thousand and One Nights, The 56
Three-toed sloth, 178–82
Thutmose I, 279
Tibareni, 123
Tiger shark, 64
Tigers, 299
Titicaca, Lake, 202–3
Tlachtli (game), 267
Tobacco addiction, in monkeys, 233–34
Tooth decay, 109
Tornadoes, 83–84
Trap-door spiders, 186–87
Treasury of the Poor (Thesaurus Pauperum), 44
Trees
  cottonwood, 166
  fossilized, 262–63
  ginkgo, 232
Triassic Period, 228–29
Triceratops, 4–6
Tropisms, 185–86
Tsunamis, and extinction of dinosaurs, 304–5
Tuatara, 93, 228–29
Turbot, 10–11
Turkey buzzard, 6–8
Tutankhamen, 141, 243–44
Typhoons, 29–33, 212
Tyrannosaurus rex, 1–6, 128, 223–24

Uganda, 118
Ultraviolet rays, 174
Undertaker honeybees, 86–88
Underwater volcanoes, 10–11
United States. See also Colonial America; *individual states*
  firestorms in, 46–48
  flu epidemic in, 69–70, 72–73
  quackery in, 16–19
  tornadoes in, 83–84
University of California, Berkeley, 6
Uranium rush, 37–40
Uranium sand houses, 18
Utah, uranium rush in, 38–39

Valles Marineris, 153, 237–38
Valley Forge, 114
Valley of the Kings, 279
Veil, bride's, 288
Venom, of king cobra, 41–42
Venus, 151–54
Verde Lake Beds, 260
Verne, Jules, 17
Vesuvius, Mount, 254
Viper (typhoon), 33
Virgil, 188
Virus, influenza, 69–74
Volcanoes
  and fossilized forests, 23
  marine, 10–11
  on Mars, 238
  Sadiman, 172, 173
  whirlwinds from, 45
Von Schlavrendorff, Baron, 254–55

Walrus, 79
War of the Worlds, 236
War stories, 252–55
Washington, George, 16, 114
Washington, Mount, 225–26
Washington State, earthquakes in, 158
Wasps, 186–87
Water, 120–22
  on Mars, 238
Water travel, 40–41
Weapons
  bees as, 253
  stones as, 59–61
Weather. *See* Meteorology
Welles, Orson, 236

West Virginia, fossils in, 145–47
Wethersfield (Connecticut), 89–90
Whales, polar bear attacks on, 79–80
Whippoorwill, 293
Whirlwinds, 45
Whistles, 155
Whiteouts, 174–75
White-tipped shark, 65
Wild geese, 294–95, 298
Wildebeests, 298
Wilhelm, Kaiser, 72
Wind
  on Mount Washington, 225
  world's windiest place, 120
Winston, Henry, 77
Wisconsin, Peshtigo Horror firestorm in, 46–48
Wollstonecraft, Mary, 34–35
Wood, J. G., 59
Woodpeckers, 166
Worker bees, 86–88
World War I, 28
  bees as weapon in, 253
  flu epidemic during and following, 69–74
World War II
  firestorms in, 46
  food poisoning in, 252–53
  parachuting in, 28
  shark attacks during, 62–64
  typhoons in, 30–33
  war stories from, 254–55
Worms, 282

Wouk, Herman, 32
Wright, Bruce, 182
Wyman, Jeffries, 219
Wyoming, blizzard in, 140–41

Xerxes the Great, 252

Yellowstone National Park, 23–24

Zola, Emile, 17
Zone-tailed hawk, 136
Zoology. *See also* Entomology; Ichthyology; Ornithology
  antelope, 110–11
  Arabian oryx, 290–93
  bison herds, 192–93
  defensive mechanisms, group, 298–300
  dog, first parachuting, 28
  dogs, senses in, 124–25
  elephant seal, 207–9
  elephants, 8–10, 206–7
  gorillas, 193–94, 264
  Great Mouse War, 106–8
  intoxication, 116–18
  kangaroos, 290
  lemmings, 20–23
  lions, 132–34
  mating rituals, 50–54
  monkeys, 232–34
  platypus, 220–21
  polar bears, 78–82, 297–98
  sloths, 178–82